GUIDE TO GRASSES of South Africa

Dedicated to my parents

'God will not seek thy race, nor will He ask thy birth: alone He will demand of thee - what hast thou done with the land I gave thee?'   (Ancient Persian Proverb)

FP VAN OUDTSHOORN

in co-operation with

W S W TROLLOPE
Department of Animal and Pasture Science, Faculty of Agriculture, University of Fort Hare

D M SCOTNEY
Soil and Irrigation Research Institute

P J McPHEE
Chief Directorate Agricultural Engineering

BRIZA Publikasies Cc

Reg. no. CK 90/11690/23

**P O Box 56569
ARCADIA, 0007**

First Afrikaans edition 1991
First English edition 1992

Text © F P van Oudtshoorn
Photographs © E B van Wyk excluding Figures 1, 2, 4, 5 and 12 kindly made available by Department of Agricultural Development
Illustrations © P D Terblanche, W Roux
Distribution maps H S Smal with Mapping-CAD
English edited by E du Plessis
Reproduction McManus Bros, Cape Town
Printed and bound by National Book Printers, Cape Town

All rights reserved. No part of this publication may be reproduced or transmitted in any form or by any means without written permission of the copyright holder.

ISBN 0 620 16539 1

CONTENTS

ACKNOWLEDGEMENTS 6
PREFACE 7
INTRODUCTION 8

1. SOIL EROSION AND CONSERVATION 11
(SCOTNEY & McPHEE)
Soil erosion in South Africa.....Causes of soil erosion ....Types of erosion.....Factors affecting erosion.....Prevention and control of erosion..... Reclamation of eroded areas.....Principles for effective conservation.

2. ECOLOGICAL AND PASTURE CONCEPTS 35
(VAN OUDTSHOORN)
Ecosystem.....Plant succession.....Sweetveld, Sourveld and Mixed veld.....Biomes of South Africa.....Optimal grazing.....Carrying capacity

3. VELD MANAGEMENT IN GRASSLAND AND SAVANNA AREAS (TROLLOPE) 45
Veld condition.....Veld management practices.....Veld management systems.....Veld management layouts.....Veld rehabilitation

4. GRAZING VALUE AND ECOLOGICAL STATUS OF GRASSES (VAN OUDTSHOORN) 57
Grazing value
Ecological status

5. MORPHOLOGY OF THE GRASS PLANT 62
(VAN OUDTSHOORN)
Vegetative parts
Reproductive parts
Identification system

6. SPECIES DESCRIPTIONS (VAN OUDTSHOORN) 72
Solitary inflorescences 73
Digitate inflorescences 123
Paniculate inflorescences 178
Inconspicuous inflorescences 279

GLOSSARY 282
INDEX TO PLANT NAMES 287
Botanical names 288
Afrikaans common names 292
English common names 297

ACKNOWLEDGEMENTS

To acknowledge the literally hundreds of people who have been directly or indirectly involved in this project over the past few years, is certainly not easy. A special word of thanks, however, to the following persons and institutions:

Eskom, Gencor, the Anglo American & De Beers Chairman's Fund and Lennons for their financial assistance. These days such vast amounts of money are spent on the conservation of endangered species, that the dilemma presently facing our soil is easily forgotten. The continued existence of man and all plants and animals depends on this soil. I would like to thank the above mentioned institutions for their insight in this regard.

Without Eben van Wyk this book would not have been possible. I thank him for his trust in me and in the project and for his professionalism.

Stephanus Smal for his involvement and especially for the hard work on processing the grazing values and ecological status data, as well as the preparation of distribution and other maps.

Pieter Terblanche, a good friend, who assisted with a large part of the project and was responsible for most of the illustrations, but whose involvement was interrupted by work on Marion Island.

Prof. Winston Trollope, Dr Derek Scotney and Peter McPhee for their willingness to compile the two important chapters on veld management and soil conservation.

Yvonne Venter and Emsie du Plessis for the translation of the manuscript and their faith and interest in the project.

Emsie du Plessis and Prof. Erdmann Baumbach for polishing our style, as well as for valuable comments on the English and Afrikaans manuscripts.

Lola van Rensburg for her patience and excellent typing.

The Curator of the National Herbarium in Pretoria for the use of distribution maps, the assistance and moral support of the staff, especially Dr Bernard de Winter, Marinda Koekemoer and Lyn Fish, as well as Wilma Roux who prepared some of the illustrations.

Proff. Guillaume Theron and Braam van Wyk, Department of Botany, University of Pretoria; Lyn Fish and dr Bernard de Winter, National Herbarium, Pretoria, for improvements to the manuscript through their valuable comments on botanical aspects.

Numerous farmers and reserve managers for access to their farms or areas under their control, as well as for the hospitality with which we have been received during photo sessions for this book.

More than a hundred grass experts who supplied valuable information on the grazing values and ecological status of the different species.

F P van Oudtshoorn,   April 1992

PREFACE

There can be no doubt that the conservation and effective utilization of our natural vegetation are of utmost importance for the sustained economic productivity of the game and stock-farming industry in South Africa. The efficient management of this natural vegetation resource requires a thorough knowledge of grasses, their habitat preferences, value as source of fodder for animal production as well as their reaction to different management practices. This guide supplies some of the information required for efficient veld management. It is my sincere wish that this book will be a useful aid through which farmers, agriculturists, nature conservation officers, students and lovers of our beautiful flora can train and enrich themselves in the wonder of nature. May this handbook contribute positively towards a better understanding and appreciation of the conditions under which different grass species grow, their value and management requirements, enabling us to employ that which our Creator has entrusted to us, to His glory to a better quality of life. May it contribute to a greater appreciation of our vegetation and may we receive the grace to arrest the deterioration of our natural vegetation resource. If we do not succeed, we shall have to account for it. My hearty congratulations to the authors of the book. Your efforts are much appreciated.

A J Aucamp (Sci. Nat.)
Head: Pasture Research Centre
Pretoria

INTRODUCTION

Many people regard grass merely as grass. However, the grass family is probably the most important plant family for man's survival and economic prosperity. What would happen if products such as meat, milk, bread, butter, sugar, cereals, wool and leather, all directly or indirectly derived from grass, were to disappear from the earth? Or even worse, what would happen if millions of tons of topsoil were lost through accelerated soil erosion, impoverishing the soil resource to such an extent that food production became virtually impossible? It becomes clear that grass is very important and that man and beast depend on it. But do we always realize this? This book is aimed at the stimulation of people's interest in grass so that their actions will show that they understand, respect and appreciate its importance with regard to food production and especially to soil conservation.

Grasses are the most widespread plants on earth, occurring from deserts to rain forests. The grass family (Poaceae) is also one of the world's largest plant families. Of some 10 000 species worldwide, 967 occur in Southern Africa; 329 of these are limited to this region and 115 are exotic. At least 200 grass species may on account of their abundance be seen as dominant grasses of South Africa. These species are treated and described for identification purposes.

This book is the first national veld guide for the identification of grasses. The colour photographs illustrating the different species, are also special. The colour and habit of the grass plant depicted on the photographs, should make the usually difficult identification process considerably easier.

For every grass, the correct scientific name and both the English and Afrikaans common names are supplied. The scientific name comprises two parts, for example *Panicum maximum*. The first part or genus name may be regarded as the 'surname' of the grass, and the second part or species name its 'first name'. Unfortunately the scientific name may sometimes change as international taxonomic research progresses and new relationships are found. Common names often lead to much confusion, because every region or even every farm sometimes has its own common name for a species. This may cause a species such as *Cynodon dactylon* to have more than twenty common names. The common names given in this book are in general use and it is hoped that they will contribute to the standardization of common names.

For a better knowledge of the structure of the grass plant, there is a section in which all morphological parts are illustrated and their functions described. The distinctive morphological differences that should be noted during identification, are furthermore given point by point.

To facilitate the identification process, an identification system, based on the type of inflorescence, has been developed for the book. The inflorescences of the different species have been grouped and these groups are then treated separately. By comparing the inflorescence of an 'unknown' species with those in its group, the grass can be identified (see Identification system in Chapter 5).

However, the identification process cannot be completed without the accompanying species descriptions which give, among others, salient morphological features of the different plant parts of the species, such as the inflorescence, spikelets, leaf blade, leaf sheath and ligule. The habitat preference as well as the biomes in which the species occurs, is supplied. Useful general information such as grazing value, ecological status as well as the grazers showing a preference for a certain species, is given in some cases.

Together with the description and photographs, there is a map showing the geographical distribution of the species in question. These maps were compiled with the aid of information obtained from the National Herbarium in Pretoria.

Although technical terms have been avoided as far as possible, it was nevertheless found necessary to explain more than 160 botanical, pasture and soil conservation terms in the glossary. Indices to scientific names, English common names and Afrikaans common names have been included for quick and easy reference.

However, there would be little sense in stimulating interest in grasses without providing more information on soil conservation, one of the main aims of the book. Soil Conservation requires the maintenance of a good veld management programme by the land user, and sections on soil conservation and veld management have therefore been included. The former deals with, among others, the causes of soil erosion, the process of soil erosion and its prevention and control. In the chapter on veld management, main points such as veld condition, veld management practices and veld management systems are discussed.

After alarming facts about the poor condition of the veld in South Africa had come to light, the National Grazing Strategy was promulgated in Parliament in 1985. Its main aim is to utilize, develop and manage natural and artificial pasture in South Africa to the greatest sustained advantage of the present generation, while the production potential is maintained to satisfy the needs and aspirations of future generations.

The optimal utilization of a natural resource cannot be effected without a phase of awareness and then training which should involve all, from the farmer to the agricultural extension officer and the environmentally aware general public. The Chinese have a saying:

'If you plan for a year, plant rice,
if you plan for a decade, plant trees
but if you plan for a lifetime, educate.'

The realization of this fact has led to the **Guide to grasses of South Africa** and it is to be hoped that the book will make a real contribution to correct veld management and soil conservation.

1 SOIL EROSION AND CONSERVATION

From the beginning of time, erosion has shaped the South African landscape. Its majestic mountain ranges, deeply incised valleys and vast open plains are largely the result of natural erosion. The agents of both water and wind have removed weathering material from sites and deposited it elsewhere to form the landscape as it is today. This slow process (**natural erosion**) has taken place under conditions of more or less equilibrium between soil formation and erosional losses.

By contrast, such equilibrium has been drastically disturbed by the activities of man which have greatly increased the rate of erosion (**accelerated erosion**). One of the prime reasons for this erosion is the removal or destruction of the protective cover of natural vegetation. In many parts there is clear evidence that soil losses far exceed the rates of natural erosion (Scotney & McPhee, 1990). This has important economic consequences for the nation and implies a loss of soil productivity over the long term.

This chapter reviews soil erosion in South Africa, especially that associated with natural grazing. Consideration is given to the erosional process, types of erosion and principles concerned with control measures.

SOIL EROSION IN SOUTH AFRICA

Despite the concerted efforts of many dedicated conservationists, the stewardship over our vital soil resources has not met the challenge (Adler, 1985). As early as 1923, the Drought Investigation Commission reported that accelerated erosion and desiccation were attributable to man's misuse of the land. In 1948, Ross reported that widespread veld deterioration and denudation were to be observed in all parts of the country.

Tidmarsh (1948) also found that in the Karoo and adjacent grassveld areas considerable deterioration of the veld had taken place over decades. Many eminent international scientists have also alluded to the serious state of veld degradation and erosion in South Africa (Robertson, 1968) (Figure 1).

Figure 1. Severe degradation caused by erosion.

Many natural features render the landscape vulnerable to erosion. High rainfall energy, steep slopes and soils of high erosion hazard favour high soil losses once the natural vegetation is destroyed. The South African soil mantle is highly diverse, complex and largely susceptible to erosion (Scotney & McPhee, 1990) (Figure 2). About 80% of the country comprises slightly weathered and calcareous soils which tend to be shallow and of unfavourable morphological properties, chemical composition and clay mineralogy. Topsoils of very low organic matter content (<5% organic carbon) characterise almost 60% of the country. Over 30% comprises excessively sandy soils (<10% clay) which are particularly vulnerable to wind erosion.

Important relationships between water and wind erosion, vegetative cover and annual rainfall have been

Figure 2. Highly erodible duplex soil.

described by Branson *et al.* (1981) and are reflected in Figure 3. These are particularly relevant to South Africa since about 60% of the country receives less than 500 mm rainfall per annum.

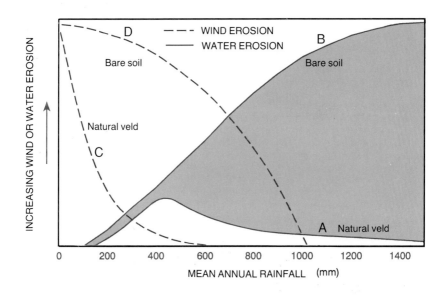

Figure 3. Relationship of water and wind erosion for bare soil (B, D) and natural veld (A, C) with increasing annual rainfall.

Curve A represents natural veld and shows water erosion increasing from arid to semi-arid conditions. Thereafter it decreases as rainfall and vegetation cover improve. Curve B reflects water erosion from bare soil. Curves C and D relate to wind erosion and reflect very different relationships. Wind erosion in arid areas, even in natural veld, can be very severe, especially where sandy soils predominate. When the amount and distribution of rainfall as well as vegetative cover improve, the hazard of wind erosion is greatly reduced, even for bare soil. Figure 3 clearly demonstrates the important role played by natural veld and the potential for preventing erosion through the application of sound veld management practices, especially in semi-arid and humid areas.

The continuing degradation of soil and veld resources (80 million ha) in South Africa led the Government to declare a National Grazing Strategy in 1985. This Strategy aims to halt resource degradation through the implementation of various research, extension and educational programmes. In motivating the Strategy, Bruwer (1986) noted that in little more than two decades the Karoo had encroached 70 km north-eastward into grassland; encroachment of woody species had rendered 3 million hectares of Bushveld useless for grazing by domestic livestock; 60 percent of the veld was in poor condition and in almost all veld types, current stocking rates were in excess of the grazing capacity. A recent survey of almost 20 000 species of plants in southern Africa has shown that well over 2 000 species are seriously threatened by extinction with as many as 36 species of the Cape Floristic Kingdom already extinct (Council for the Environment, 1989). An example of ecosystem destruction is reflected in the high percentage (57%) of wetlands damaged by gully erosion in the Mfolozi catchment of Natal (Begg, 1988).

Accurate estimates of water and wind erosion in South Africa are not yet available, but according to Adler (1985) sediment losses for large catchments are of the order of 300 million ton per annum or an average of 3 ton/ha/annum for the country. However, this estimate is based on the sediment carried by major rivers and does not reflect the gross erosion over the landscape. Assuming a sediment delivery ratio of 0,06 for larger catchments, actual soil losses could be as high as 50 ton/ha/annum in some areas.

Braune & Looser (1989) suggest rates of natural erosion in South Africa of between 0,5 and 1,4 cm per 1 000 years. They give the rate of accelerated erosion as 12 cm per 1 000 years; this implies that current soil losses may be more than 20 times the rate of soil formation. This is very much higher than the world average of 3 cm per 1 000 years. Adler (1985) also estimated annual soil losses per capita to be almost 20 times the world average.

The economic consequences of erosion are stressed by Huntley *et al.* (1989) who claim that the almost 3 million hectares rendered unusable as a result of severe erosion, had an asset value of R1 500 million. They also give the value of plant nutrients in sediment carried annually to the sea by rivers, as over R1 000 million.

Damage caused by wind erosion is not fully appreciated in South Africa nor have losses been accurately estimated. Under certain cropping systems soil losses may reach some 60 ton/ha/annum, and even under natural veld conditions exceptionally high losses may result from a single windstorm (Scotney & McPhee 1990).

The extent and cost of off-site damages caused by erosion are most important. These are often far higher and of greater economic consequences than on-site damages. For example, the Demoina flood of 1984 deposited some 34 million tons of sediment on the valuable Mfolozi flats, causing untold damage to agriculture, tourism and recreation (Looser, 1985). Sediment seriously reduces water quality and in some catchments has drastically reduced the storage capacity of dams in a very short space of time. In the case of the Welbedacht Dam on the Caledon River, 32% of its capacity was lost in the first three years (Scotney & McPhee, 1990). Braune & Looser (1989) recently estimated that the cost of such losses in South Africa may far exceed R50 million per annum. All land users should be fully aware of the impact of such off-site damages.

No matter how tentative current estimates might be, the seriousness of soil erosion in South Africa is beyond question.

Legislation and financial assistance

All owners of agricultural land are subject to the provisions of the Conservation of Agricultural Resources Act, No 43 of 1983. The stated objectives of this Act are to 'provide for the conservation of the natural agricultural resources of the Republic by the maintenance of production potential of the land, by combating and prevention of erosion and weakening or destruction of water sources, and by the protection of the vegetation and the combating of weeds and invader plants.' The onus rests on each landowner to be fully *au fait* with the provision of this Act and all other legislation (e.g. Environmental Conservation Act No 70 of 1989) of relevance to their farming activities and environmental management.

To achieve the objectives of the Act, the Minister may prescribe control measures relating to such matters as the utilization and protection of cultivated land, the natural vegetation, veld which has burned as well as vleis, marshes, water courses and other water sources. Measures relating to the grazing capacity of veld, the maximum number and the kind of animals which may be kept on the veld, the control of weeds and invader plants, the reclamation of eroded land and the construction and maintenance of soil conservation works, may also be prescribed.

Many of the aspects of this Act are of particular relevance to land-owners utilizing natural veld and can have far-reaching consequences for transgressors in the form of severe penalties. The Act also makes provision for financial assistance in the form of subsidies for the construction of soil conservation works and for achievement of other conservation goals, provided these are undertaken in accordance with prescribed specifications. Financial aid rendered to farmers under this Act and many other financial assistance schemes, amounts to millions of rand.

CAUSES OF SOIL EROSION

Soil erosion represents a loss of capital not generally recognized in the value of farmland. It is seldom appreciated that the cost of soil degradation will ultimately fall on the shoulders of future generations. Without considering the actual process of erosion, the main causes can be viewed from both a general and a more detailed level.

General causes

Increasing pressure on land resources. Rapid population growth has resulted in the land available per capita dropping from 5,5 ha to 3,2 ha in only 20 years (Adler, 1985). Over the past 80 years the area cultivated in South Africa has increased from 4 million ha to almost 13 million ha. An even greater rate of expansion applies to the area planted to maize which has an unfavourable surface cover. Tractor numbers increased rapidly over a period of 60 years to peak at almost 240 000 units. Sheep numbers increased rapidly to peak at 48 million head in the early 1930s, a period associated with much soil degradation. Cattle numbers have doubled since the early 1900s but have tended to stabilize over the past two decades. A growing demand for land for purposes other than agriculture, such as urban development, industries, infrastructure and afforestation, has also led to considerable losses of high potential farmland. Furthermore, much intensification has taken place on marginal land without adequate conservation. Pressures such as these have led to severe degradation of the soil resource.

Economic influences. In many instances economic influences have led to over-exploitation. Strong profit motives, unrealistic commodity and land prices and uneconomic farm sizes have led to the application of many non-adapted land-use systems, with serious consequences.

Incorrect farming systems. South Africa's low and erratic rainfall, regular droughts, shallow and erodible soils and rugged and rocky land demand that careful attention be given to the natural resource base in selecting land-use alternatives. Failure to apply appropriate systems and environmental management has led to much degradation.

Lack of knowledge and expertise. Manipulation of natural ecosystems inevitably leads to a simplification of the biotic components, often accompanied by severe degradation. Prevention of this demands management of high order and clear understanding of ecological processes. The imbalance of these two factors - biotic simplification and management expertise - has been referred to as the 'scourge of agriculture' (Booysen, 1980). Many land managers are not adequately trained and perceive the process and problem of degradation incorrectly. There is little evidence of widespread support for a strong land care ethic.

Inadequate law enforcement. Conservation effort requires the application of effective legislation and regulatory control. Experience suggests limitations in past attempts to bring blatant transgressors to book. Furthermore, many of the financial schemes available to farmers have not resulted in the desired improvement in conservation practise.

Causes at the detailed level

Planning. Despite concerted efforts over many years, farm planning as a basis for sound conservation has not proved highly effective on a national scale. Over-emphasis has been given to physical planning, to the detriment of important biological and economic considerations. In many instances the selection of enterprises has not been in accordance with the dictates of the natural resources. Poor siting of fences, roads and paths, inadequate provision of stock watering and ineffective conservation works also contributed to the problem of soil erosion.

Management. Lack of sound farm management and attention to detail are important causes of resource degradation at farm level. For instance, there has been a failure to proactively plan for drought, adjust stocking rates to grazing capacities, use fire as an effective management tool, prevent indiscriminate deforestation and apply appropriate and modern technology.

These are some of the basic causes that have led to accelerated erosion and widespread degradation of the natural veld.

TYPES OF EROSION

Water erosion

Erosion begins when a raindrop moving at a terminal velocity of about 20 km/h strikes a bare and unprotected soil surface (Figure 4). The two main processes are the detachment of soil material from the surface by drop impact and/or runoff shear and the transport of sediment by drop splash and/or overland flow.

The mass of a raindrop 2,5 mm in diameter is approximately 12 000 times greater than the mass of a soil particle of 0,1 mm diameter. Thus the kinetic energy of the impact may move a particle of this size over a considerable distance. This initial form of erosion is known as **splash erosion**. The pounding action of raindrops on the surface also causes a seal to form, resulting in reduced infiltration and increased runoff.

Once the infiltration capacity of the soil is exceeded by the rainfall rate, runoff begins to concentrate in small channels, soil particles are transported downslope and scouring of the soil surface begins. This produces the next type of erosion known as **rill erosion**. Splash erosion with a minimum degree

Figure 4. Splash erosion is caused by the impact of a raindrop on the bare soil surface.

Figure 5. Flood plain dissected with gully erosion.

of rill erosion is commonly known as **sheet erosion**. As the process continues, runoff in the rills becomes deeper and wider to form the most spectacular type of erosion, known as **gully erosion** (Figure 5).

Splash, rill and gully erosion are the main forms of erosion with which the land user has to contend, but other forms of erosion may also be significant. These include:

Terracette erosion. This is the result of a natural process which takes place on steep slopes when mass downslope movement of soil results in a series of steps of varying width. The vertical faces of such steps are usually bare of vegetation. Terracette erosion is a complex natural phenomenon which may be aggravated by the tramping of cattle and sheep.

Streambank erosion. This type follows the splash-rill-gully erosion sequence. Fast flowing water undermines the outside bank of a channel, causing its eventual collapse and this results in the widening of the stream or river.

Tunnel erosion. This type of erosion occurs in veld adjacent to gullies, in depressions or in earth embankments. It is characterised by the appearance of vertical or horizontal tunnels which may ultimately collapse. Soils with high shrink-swell potential and a high sodium content are particularly prone to tunnel erosion.

Wind erosion

During the process of wind erosion, soil particles are transported by three mechanisms, namely creep (rolling along the ground surface), saltation (bouncing into the air and landing some distance away) and suspension (taken up into the moving air and blown over large distances). Exceptionally high losses have been experienced from ploughed fields (Matthee & Van Schalkwyk, 1984). Even under natural vegetation in low rainfall areas, soil losses caused by wind erosion may be extremely high. Fine textured top soils ($<10\%$ sand) with sparse cover and under dry conditions, are particularly vulnerable to wind erosion when critical wind velocities (20 km/h) are exceeded.

Deposition

Soil erosion implies movement of sediment over the landscape. Major shifts in soil fertility result from such movements and in severe cases, during floods for example, large deposits may cause untold damage to downstream properties (Braune & Looser, 1989; Looser, 1985). However, much depends on the textural characteristics of the sediment. Sandy materials are generally of less benefit than medium to fine materials of higher fertility. The latter are more easily reclaimed by re-vegetation.

FACTORS AFFECTING EROSION

Water erosion

An erosion equation, used for predicting average long term soil losses by sheet erosion (splash and minor rill erosion), developed by Wischmeier & Smith (1978) and called the Universal Soil Loss Equation (USLE), deserves mention since it enumerates the various causal and preventive factors in erosion. Figure 6 represents a simplification of the relative importance of selected factors contributing to sheet erosion (Roose, 1981). These factors include crop cover and management, slope steepness and length, soil erodibility and tillage direction. Rainfall erosivity is excluded on account of its variability.

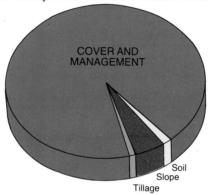

Figure 6. The relative influence of certain factors affecting water erosion.

It is obvious that the cover and management factor is many times more important than the other factors and that the destruction of veld and crop cover will greatly affect erosion. The following factors contribute directly towards the vulnerability of soil to water erosion:

Rainfall erosivity. Rainfall erosivity is the potential of rainfall to cause erosion and combines rainfall intensity with the energy of raindrops. Local erosivity values have been developed and are available in map form (Smithen, 1981). Values range from 50 units or less along the west coast to over 500 units in the Eastern Transvaal.

Soil erodibility. The basic erodibility of a soil is the soil loss per unit of rainfall erosivity. Soil erodibility can be quantified by measuring soil losses from bare runoff plots subjected to natural or simulated rainstorms. Rainfall simulators have been used in die RSA for this purpose. Soil erodibility can also be predicted by means of the nomograph developed by Wischmeier & Smith (1978).

The initial phase of soil erosion is restricted to the top few millimetres of the soil surface. Some soils, especially in semi-arid areas, are particularly prone to crust formation as a result of mechanical and chemical processes. These crusts seal the surface, reduce infiltration, increase runoff and may interfere with seedling emergence.

Slope length and steepness. Slope length (defined as the distance between a point where runoff is initiated to another point down the slope where runoff is intercepted) and steepness are combined in a topographical factor for the purpose of soil loss prediction. Steepness has greater influence than slope length. For instance, for a 30 m slope length, an increase in slope from 2 to 8 % will result in a five-fold increase in soil loss. By comparison, an increase in slope length from 30—120 m on a 5% slope will result in only a two-fold increase in soil loss.

Cover. Agricultural intensification, unfavourable economic conditions, tightening economy, serious effects of drought and sub-optimal management have resulted in the severe degradation of the natural veld. With both tilled and undisturbed soils, it has been shown that mulch and basal veld cover respectively are extremely effective in controlling erosion, more so than canopy cover (McPhee & Smithen, 1984). These dissipate the rainfall erosivity at ground level. Cover on the ground also obstructs and retards the flow of runoff. Mulch cover also prevents a seal from forming on the surface. The percentage reduction in soil losses from bare ground due to cover on both tilled soils and natural veld in the RSA and in the USA is shown in Figure 7.

Snyman & Van Rensburg (1986) have quantified the effect of basal cover and plant successional stage on soil loss and runoff by two rainfall simulator storms in natural veld on a Valsrivier soil form. Selected and summarized results are given in Figure 8 which shows the marked effect of basal cover on slopes ranging from 2—6%.

Management. The effect of cover is so dominant in erosion control, that it is imperative that the farmer carefully evaluate the effect of his management on veld condition. A wide choice of management options are available for controlling erosion, improving botanical composition and veld condition and thereby optimising production. Where the veld has been severely damaged, total withdrawal from grazing may be necessary. Strict control of stocking rates, appropriate grazing systems, judicious burning regimes and even the application of radical veld improvement techniques, are some options available to the farmer.

Runoff intensity. The potential soil loss from untilled soils is far less (45%) than that from tilled soils (Wischmeier & Smith, 1978). However, there is no reason for complacency since grazing land is likely to be steeper and prone to higher runoff intensities than tilled land. Natural veld is normally not protected by erosion control works such as contourbanks. The distance over which runoff flows in natural veld, is usually greater than in cultivated

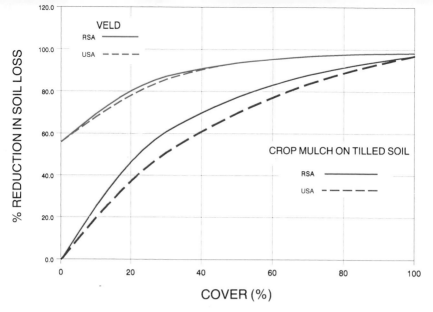

Figure 7. The percentage reduction in bare plot soil losses due to cover on tilled soil and veld for R.S.A. and U.S.A.

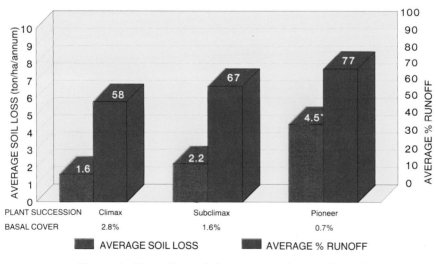

Figure 8. The effect of plant succession and basal cover on soil loss and runoff in natural veld (Snyman & Van Rensburg, 1986).

lands. In addition, soils in grazing land are likely to be less well-drained, shallower and more vulnerable than soils in tilled land.

On a soil of moderately low runoff potential and with a slope of 8%, the runoff intensity from veld in poor condition in a 100 hectare Highveld catchment area, could be three times greater than that from veld in good condition (Matthee & Van Schalkwyk, 1984). This would increase the erosion capacity of runoff by a factor of 900%

Wind erosion

Wind erosion losses are highest in areas where soils are dry and sandy. Wind velocity, average monthly precipitation (wet soils do not easily erode) and temperature determine the susceptibility of soils to wind erosion. Soil erodability is based largely on the size of aggregates in the top 2,5 cm of the soil profile. Such losses are also affected by surface roughness caused by ridges due to tillage. Ridges between 5 and 10 cm high, with a spacing of four times the height, have been found to be effective in reducing erosion (Armbrust *et al.*, 1964). The length of field in the direction of the prevailing wind can also have a considerable effect on erosion losses. As with water erosion, the quantity, type and orientation (flat or standing) of the vegetative cover are the most important factors.

PREVENTION AND CONTROL OF EROSION

Prevention and control of erosion are feasible provided certain principles are adhered to and management is of high order.

Optimal resource utilization

The concept of optimal resource utilization is entrenched in agricultural policy now widely accepted in the RSA. It requires that farm practice should:
- be in harmony with the natural environmental factors,
- not be achieved at the cost of the natural resources, and
- have a sound economic base.

Achievement of the above points should be the aim of each soil user.

Soil life concept

The balance between rate of soil formation and soil loss determines the 'life' expectancy of a soil. Little is known about formation rates of soils in South Africa, but it may be assumed that soils form at an extremely slow rate (e.g. 0,31 ton/ha/annum) (Scotney & McPhee, 1990). Soil losses from farmland, even under natural vegetation, usually far exceed formation rates. In such cases soil life is drastically reduced and productivity becomes impaired over time. Severe and even permanent damage occurs when soil depth is markedly reduced by erosion.

Assuming the above rate of formation, soil life can be calculated for soils of varying depth and for different erosion rates. This approach is illustrated in Figure 9. The soil life concept emphasizes the important need for restricting soil losses to acceptable levels.

Soil loss tolerance values

Allowable soil losses or tolerance values are those that will allow productivity to be maintained and provide land managers with important targets for their conservation efforts. Such values differ for various soils and for arable and non-arable situations. While no serious attempt has yet been made to en-

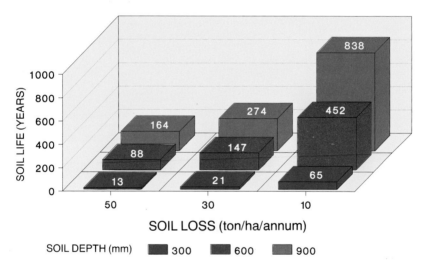

Figure 9. Diagrammatic presentation of the soil life concept.

trench tolerance values in the local conservation ethic, the concept of restricting erosion rates to a tolerance level is sound. Guideline tolerance values for arable and non-arable situations are of the order of 5—10 ton/ha/annum and 0,5—1,0 ton/ha/annum respectively (Wischmeier & Smith, 1978).

Farm planning

The most essential requirement for successful erosion control, is that each farm is planned in accordance with the dictates of the natural resources. It is not intended to describe the principles and procedures for farm planning, since these are readily available from advisory services. However, logical steps should be followed and include the phases of **survey**, **assessment**, **planning**, **implementation** and **evaluation**. Physical and economic aspects are of equal importance and should be given careful consideration in developing a farm plan appropriate to the natural, managerial and financial resources. The skills and preferences of the farmer can greatly influence the final choice of plan. The Conservation of Agricultural Resources Act (No. 43 of 1983) also requires that all farms be appropriately planned before financial aid is made available to farmers.

Runoff control planning

The impetus for current runoff control planning is provided by the Conservation of Agricultural Resources Act which requires each landowner to ensure that his land is protected from erosion. Common law also implies that no floodwater may reach a farm boundary in an unnatural manner or in a more concentrated form than would occur under natural conditions. The concept of natural responsibility acceptance implies that an owner of lower-lying land must take responsibility for receiving and disposing of runoff reaching his land *ex natura*. A basic set of principles has been proposed by Van Staden (1980), enabling a runoff control plan to be developed that is acceptable to both the farmer and the adviser. These require that:
- natural watershed boundaries are respected,
- natural watercourses be recognized as such and should not be cultivated,
- the planning of runoff disposal should start at the highest point in the catchment unless topographic conditions dictate otherwise, and
- the construction of the waterways should begin at the lowest point in the catchment.

These principles apply equally well to natural veld and arable land. A runoff control plan should specify all conservation works such as stormwater drains, waterways, contourbanks and diversion structures as well as a variety of other structures that may be necessary for controlling runoff and erosion. Contourbanks are seldom constructed in veld and therefore cover assumes great importance. It is the only practical way of reducing runoff intensity and velocity and preventing erosion.

Conservation works and measures

The farmer has many options in selecting specific erosion control measures. The principle to be applied, is that each situation should be treated according to its individual conservation needs. The design and construction of many types of works such as contourbanks, waterways and stonepacks have been adequately described by Matthee & Van Schalkwyk (1984), and are recommended for study. Measures such as rotational grazing and resting, withdrawal of eroded areas and stocking rates are discussed in Chapter 3.

Streambank protection

Streambank erosion is a natural phenomenon, but is aggravated by increased runoff from catchment areas and the destruction or riverine vegetation. Control measures should be aimed at protecting such vegetation, retarding flow velocity or protecting the streambank itself. Pole jacks (concrete or wooden), as illustrated in Figure 10, are widely used to form barriers which retard flow velocity, collect flotsam in flood water and promote sediment deposition. The common reed (*Phragmites australis*) can then be established to help stabilize the channel. Groynes consisting of gabions can also be used to deflect the stream away from the eroding bank. In swiftly flowing water or at sharp bends in the river, banks can be protected by some form of flexible lining made of brushwood (pegged down), timber logs, stone riprap, gabions, articulated concrete slabbing or Reno mattresses.

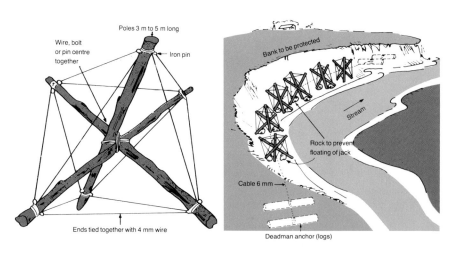

Figure 10. Streambank protection by means of pole jacks.

Wind erosion control

Effective control of wind erosion in veld can be achieved only by ensuring a cover of vigorously growing vegetation. In some cases windbreaks may prove useful. For planning purposes, it may be assumed that a windbreak of trees will shelter five times the height of the trees on the windward side and 15 times the height of the trees on the leeward side.

RECLAMATION OF ERODED AREAS

Valuable information concerning the reclamation of eroded areas is available from standard reference works (Pentz, 1955; Roberts, 1981; Scott, 1981 and Matthee & Van Schalkwyk, 1984). Careful assessment of each site is needed before reclamation commences, since the method employed will be determined by the degree, extent and causes of denudation, the urgency for reclamation and the financial implications. Two approaches are normally followed, namely increasing vegetation cover through improved veld management (especially resting) where the potential for recovery is high, and applying special treatments such as mechanical soil disturbance (e.g. ripping) and re-seeding where areas are severely degraded.

Treatment of bare areas

In order to reclaim and re-vegetate such areas, it is necessary to eliminate the major causes and improve the soil moisture status before re-establishing appropriate vegetation. Where soil crusts are formed by mechanical processes, the application of a mulch or suitable cover will prove highly effective. However, where chemical processes are concerned, the application of an appropriate ameliorant (e.g. phosphogypsum) to the soil surface is likely to have best results, since mechanical disturbance is unlikely to be effective over the long term. Soil structure and aggregate stability may also be improved by the application of soil conditioners such as organic polymers.

Many alternatives are available for re-vegetating bare areas. Single stone packs on the contour with the area in between ripped and re-seeded with local grass species, have proved successful in many parts. Another method is to use a basin plough which creates a series of small basins in the soil surface. These basins are capable of storing a considerable amount of rainfall and will promote infiltration. Virtually all runoff may be eliminated on flat land by using this method. However, on sloping land, basin tillage may result in large soil losses if the basins are overtopped.

Where bush encroachment is associated with the sparse grass cover, trees should be cleared and the brush packed over the area. Protection of the area by this means is often highly successful. In time the brush decomposes, and then the area can be lightly grazed once again. Resting of such areas during the growing season is highly desirable.

Gully-eroded flood plains

All forms of reclamation aim to restore the eroded area to more or less its original state. This requires careful planning, astute design and constant maintenance of works. It should be fully appreciated that such reclamation may be excessively costly and that the chances of complete restoration are slight. Small and inexpensive stonepacks or even brush may often prove more successful than large costly structures.

A full description of the various structures and procedures available is beyond the scope of this discussion; the reader is referred to the publications of Matthee & Van Schalkwyk (1984) and Scott (1981). It is of paramount importance, however, to appreciate that few structures, large or small, will succeed unless their use is accompanied by improved veld management throughout the catchment. It may be necessary to fence off and protect the eroded area containing the structures, but such areas need constant monitoring. Examples of structures differing widely in cost are shown in Figures 11 and 12.

Figure 11. Check dam of packed stone.

Figure 12. Large multi-arch dam to trap sediment.

Grasses in erosion control

A dense sward of grass or other adapted plants is the best means of reclaiming eroded areas and protecting conservation works such as waterways and embankments. For example, much success with gully reclamation can be ascribed to the establishment of the common reed (*Phragmites australis*) on moist sediment. Grasses such as couch grass (*Cynodon dactylon*) and kikuyu (*Pennisetum clandestinum*) have wide adaptability and can be strongly recommended for many purposes.

Where re-seeding or vegetative establishment is justified, species selection will be determined by the climatic and edaphic conditions (including fertility status) and the envisaged use and management of the area. In general, it is preferable to use indigenous perennial species. The use of awned seeds also has decided advantages when bare areas have to be re-seeded. Rhizomatous and stoloniferous grasses are recommended for disturbed soil (e.g. embankments), steep slopes and areas where runoff velocities are high. A competent pasture specialist should be consulted before initiating a re-vegetation programme, since species selection, seeding mixtures, planting procedure, fertilization and the after-care programme are all important.

Several examples of grasses and other plants suitable for erosion control in summer rainfall areas, are presented in Table 1.

TABLE 1. Different grass species suitable for erosion control (Scott, 1955; Walsh, 1968; Tainton *et al.*, 1976; Cave, 1990).

BOTANICAL AND COMMON NAMES (* Indicates exotic species)	GROWTH HABIT AND PROPAGATION							HABITAT										
								Rainfall			Temp.		Soil type					
	Annual	Perennial	Tufted	Stolons	Rhizomes	Seed	Roots	High	Average	Low	Warm	Cold	Deep	Shallow	Sand	Loam	Clay	Wet
Acroceras macrum Nile grass		X			X	X	X	X	X		X	X	X			X	X	X
Aristida adscensionis Annual Three-awn	X		X			X			X	X	X	X	X	X	X	X		
Cenchrus ciliaris Blue buffalo grass		X	X			X			X	X	X		X	X	X	X		X
Chloris gayana Rhodes grass		X	X	X		X			X		X		X		X	X		X
Chloris virgata Feathered Chloris	X		X			X		X	X	X	X	X	X	X			X	X
**Cynodon aethiopicus* Star grass		X		X		X	X		X		X		X		X	X		
Cynodon dactylon Couch grass		X		X	X	X	X	X	X	X	X		X	X	X	X	X	
Digitaria eriantha Finger grass		X	X			X		X	X		X		X			X	X	X
Ehrharta villosa Dune Ehrharta		X			X	X		X	X		X		X		X			
Enneapogon cenchroides Nine-awned grass	X		X			X			X	X	X			X	X	X	X	
Eragrostis curvula Weeping love grass		X	X			X			X		X	X	X	X	X	X	X	
Eragrostis lehmanniana Lehmann's love grass		X	X			X			X	X	X		X	X	X	X		
Eragrostis superba Sawtooth love grass		X	X			X			X	X	X				X	X	X	
**Eragrostis tef* Teff	X		X			X		X	X		X	X	X	X	X	X	X	X
Hyparrhenia hirta Common thatching grass		X	X			X		X	X	X	X	X	X	X		X	X	
**Paspalum dilatatum* Common Paspalum		X	X			X		X	X		X		X			X	X	X
**Pennisetum clandestinum* Kikuyu		X		X	X	X	X	X	X		X	X	X			X	X	
**Pennisetum purpereum* Elephant grass		X	X			X		X	X		X		X			X	X	
Phragmites australis Common reed		X			X	X	X	X	X	X	X	X	X		X	X	X	X
Sporobolus fimbriatus Bushveld Dropseed		X	X		X	X			X	X	X		X			X	X	
Stipagrostis uniplumis Silky bushman grass		X	X			X			X	X	X		X		X	X		
Thinopyron distichum Coastal wheat grass		X			X	X		X	X		X		X		X			

Fertile	Acid	Alkaline	Drought	Low pH	Salt	Floods	Saltspray	Waterways	Disturbed	Undisturbed	Dongas	Flood plains	Steep slopes	Spoils	Dunes	Wind breaks	Grazing	BOTANICAL AND COMMON NAMES (* Indicates exotic species)
X	X				X			X					X				X	*Acroceras macrum* — Nile grass
		X	X						X	X	X		X	X				*Aristida adscensionis* — Annual Three-awn
X		X	X						X	X							X	*Cenchrus ciliaris* — Blue buffalo grass
X	X								X	X							X	**Chloris gayana* — Rhodes grass
		X	X	X					X	X		X		X				*Chloris virgata* — Feathered Chloris
							X	X		X	X	X					X	**Cynodon aethiopicus* — Star grass
X	X	X	X		X	X		X		X	X	X	X	X			X	*Cynodon dactylon* — Couch grass
X	X			X	X				X	X				X			X	*Digitaria eriantha* — Finger grass
		X			X	X									X			*Ehrharta villosa* — Dune ehrharta
	X	X	X						X	X				X				*Enneapogon cenchroides* — Nine-awned grass
X	X		X					X	X				X	X	X		X	*Eragrostis curvula* — Weeping love grass
		X							X	X							X	*Eragrostis lehmanniana* — Lehmann's love grass
		X	X						X	X	X		X	X			X	*Eragrostis superba* — Sawtooth love grass
			X						X	X	X	X		X	X		X	**Eragrostis tef* — Teff
	X		X						X	X			X		X		X	*Hyparrhenia hirta* — Common thatching grass
X	X		X		X			X	X				X	X			X	**Paspalum dilatatum* — Common Paspalum
X	X		X					X	X			X		X	X		X	**Pennisetum clandestinum* — Kikuyu
X	X		X						X							X		**Pennisetum purpereum* — Elephant grass
X	X	X			X							X						*Phragmites australis* — Common reed
X			X									X					X	*Sporobolus fimbriatus* — Bushveld dropseed
		X	X						X	X			X				X	*Stipagrostis uniplumis* — Silky bushman grass
		X	X		X		X								X			*Thinopyron distichum* — Coastal wheat grass

PRINCIPLES FOR EFFECTIVE CONSERVATION

A number of important principles should be kept in mind when planning a conservation programme. These include the following:

Prevention is better than cure. Long term maintenance of productivity demands that soil be managed as a non-renewable resource. Application of sustainable and profitable farming systems which ensure resource protection, is the best means of providing for the needs of future generations. Since the reclamation of severely degraded land is costly, seldom economic and generally not feasible on the human time scale, the aim should be prevention rather than cure.

Knowledge of the natural resource base. Sound land-use planning demands a thorough knowledge of the agricultural natural resources and the associated ecological processes. Such knowledge is essential for all land managers and allows for correct land-use decisions to be made. Awareness of the potential, limitations and management needs of such resources permits the adoption of an optimal resource utilization approach.

Vegetation cover is the key to land protection. Destruction or weakening of vegetation cover inevitably leads to accelerated erosion. It is therefore imperative to maintain at least a minimum amount of vegetation cover, especially during periods of drought. A protective cover of natural vegetation, planted pasture, tree, brush, mulch or trash, is the key to land protection.

Selection of appropriate farming systems and enterprises. Systems of farming and enterprises that are adapted to the resource base, are generally the safest and most profitable. Application of non-adapted enterprises usually demands large inputs of capital and specialized management. To be effective, farm planning must take account of this important principle.

Soil loss tolerance as an objective. Soil conservation requires clear objectives to be set. Striving to contain soil losses to accepted tolerance levels, represents such an objective. The land user often has the choice of several management strategies in attempting to meet specific soil tolerances. Final selection of such strategies will be determined by the specific situation, the available resources and the economic consequences of such actions.

Assessment and monitoring of conservation status. It behoves each land user to assess the extent of resource degradation on his property and to institute a regular monitoring system for this purpose. Without such steps it is unlikely that timeous evasive action will be possible or that conservation works will be adequately maintained. It should be appreciated that much damage and loss of fertility takes place before accelerated erosion can be seen.

Rehabilitation is not necessarily the correct choice. Reclamation of severely degraded areas is seldom economical, although there may be other important benefits to be reaped. For this reason, a comprehensive assessment, including a cost-benefit analysis, should be made before large-scale reclamation schemes are initiated. It is also seldom adequate to simply remove livestock from areas, without additional conservation measures.

Land degradation and a new land ethic. Past conservation programmes have been largely ineffective in persuading farmers to apply practices that do not result in immediate economic gain. Over-emphasis has been given to physical and technical problems, without sufficient attention directed at economic and social aspects. A new approach, or ethic, in which concern for the land becomes a matter of self interest for every individual, is urgently needed.

Conservation is a process of long duration. While severe erosion may take place in a single storm, reclamation is a process of long duration. Patience and sustained effort on the part of the land user will be necessary.

Consideration to off-site damages. Off-site damages caused by erosion are often more important than on-site damages. Land users should not pursue their own aims in isolation and with little concern for the effects of their actions on others, especially those situated downstream in the catchment.

This review of soil erosion and conservation in South Africa enables the following conclusions to be drawn:

- Degradation of the natural resources by over-exploitation remains a serious threat to present and future generations.
- Land users will need to strive to improve their understanding and management of the natural resource base.
- A vigorous vegetation cover on the land is the most effective form of soil protection.
- An appropriate balance between effective conservation and economic development is essential for sustained agricultural production.
- Land users should give far greater emphasis to off-site damages, especially those resulting from water and wind erosion.
- A new land care ethic is urgently needed so that effective resource conservation and environmental protection are achieved on a country-wide basis.

REFERENCES

ADLER, E.D. 1985. **Soil conservation in South Africa**. Department of Agriculture and Water Supply. Government Printer, Pretoria.
ARMBRUST, D.V, CHEPIL, W.S. & SIDDOWAY, F.H. 1964. Effect of ridges and erosion of soil by wind. **Soil Sci. Amer. Proc.** 28: 557—560.

BEGG, G.W. 1988. The wetlands of Natal (Part 2). **The distribution, extent and status of wetlands in the Mfolozi Catchment.** Natal Town and Regional Planning Report, Vol. 17. Pietermaritzburg.
BOOYSEN, P. DE V. 1980. The importance of the ecological view of agriculture. **Proc. Grassl. Soc. South Africa** 15: 7—11.
BRANSON, F.A., GIFFORD, G.F., RENARD, K.G. & HADLEY, R.F. 1981. **Rangeland hydrology.** Kendall/Hunt Publishing Co., Toronto.
BRAUNE, E. & LOOSER, U. 1989. **Cost impacts of sediment in South African rivers.** Sediment and Environment Symposium, Baltimore.
BRUWER, J.J. 1986. **Die nasionale weidingstrategie in perspektief AKW Senario.** Dept. of Agriculture and Water Supply. Unpublished.
CAVE, A.N. 1990. **Grassing guidelines.** Dept. of Water Affairs. Unpublished.
COUNCIL FOR THE ENVIRONMENT. 1989. **An approach to a national environmental policy and strategy for South Africa.**
HUNTLEY, B., SIEGFRIED, R. & SUNTER, C. 1989. **South African environments into the 21st century.** Human & Rousseau, Cape Town.
LOOSER, U. (ed.) 1985. **Sediment problems in the Mfolozi catchment: assessment and research requirements.** Dept. of Water Affairs, HRI Report, Pretoria.
MATTHEE, J.F. LA G. & VAN SCHALKWYK, C.J. 1984. **A primer on soil conservation.** Dept. of Agriculture. Bulletin 339. Pretoria.
McPHEE, P.J. & SMITHEN, A.A. 1984. Application of the USLE in the RSA. **Agric. Eng. in South Africa** 18(1): 5—13.
PENTZ, J.A. 1955. Grasses in soil and water conservation. In D. Meredith (ed.), **The grasses and pastures of South Africa.** Central News Agency, Parow. pp. 712—723.
ROBERTS, B.R. 1981. Sweet and mixed grassveld. In N.M. Tainton (ed.), **Veld and pasture management in South Africa.** Shuter & Shooter, Pietermaritzburg.
ROBERTSON, T.C. 1968. **Soil is life.** Cape and Transvaal Printers, Cape Town.
ROOSE, E.J. 1981. **Soil loss estimation - recommendations.** Internal Report, Division Agricultural Engineering, Dept. of Agricultural Technical Services, Pretoria.
ROSS, J.C. 1948. Land utilization and soil conservation in the Union of South Africa. **Bulletin Division. Soil Con. and Ext.** Dept. Agriculture, Pretoria.
SCOTNEY, D.M. & McPHEE, P.J. 1990. **The dilemma of our soil resources.** Proceedings of National Veld Trust Conference, Pretoria.
SCOTT, J.D. 1955. Pasture plants for special purposes. In D. Meredith (ed.), **The grasses and pastures of South Africa.** Central News Agency, Parow.
SCOTT, J.D. 1981. Soil erosion, its causes and its prevention. In N.M. Tainton (ed.), **Veld and pasture management in South Africa.** Shuter & Shooter, Pietermaritzburg.
SMITHEN, A.A. 1981. **Characteristics of rainfall erosivity in South Africa.** M.Sc. thesis, University of Natal.
SNYMAN, H.A. & VAN RENSBURG, W.J.L. 1986. Effect of slope and plant cover on runoff, soil loss and water use efficiency of natural veld. **J. Grassl. Soc. South Africa** 3: 153—158.
TAINTON, N.M., BRANSBY, D.I. & BOOYSEN, P. DE V. 1976. **Common veld and pasture grasses of Natal.** The Natal Witness, Pietermaritzburg.
TIDMARSH, C.E. 1948. Conservation problems of the Karoo. **Farming in South Africa** 23: 519—530.
VAN STADEN, H.J. 1980. Planning the controlled disposal of runoff. **Agric. Eng. in South Africa** 14(1): B19—B25
WALSH, B.N. 1968. **Some notes on the incidence and control of driftsands along the Caledon, Bredasdorp and Riversdale coastline of South Africa.** Dept. Forestry, Bulletin 44. Pretoria.
WISCHMEIER, W.M. & SMITH D.D. 1978. Predicting rainfall erosion losses, a guide to conservation planning. **USDA Agric. Handbook** 537: 58.

2 ECOLOGICAL AND PASTURE CONCEPTS

ECOSYSTEM

An ecosystem consists of a biotic (living) and an abiotic (non-living) component occurring together in a specific area and adapted to each other by means of balanced interaction. The biotic component, of which grasses form an important part, includes all living organisms (from elephants to bacteria), while aspects such as rainfall, sunlight and soil form part of the abiotic component. The size of an ecosystem varies, and it can be as large as the Namib Desert or as small as a pond. There are no clear boundaries around an ecosystem and mutual interaction between ecosystems takes place constantly. The migration of animals from one ecosystem to another is an example of this type of interactions. Natural ecosystems can maintain themselves and have the ability to recover after natural disturbances such as drought and floods. With the high world population and increasing pressure on ecosystems for housing and food production, all ecosystems are artificially disturbed to a greater or lesser extent. Some ecosystems, such as residential areas and cultivated lands, have been disturbed to such an extent that little or no interaction occurs any more. Most ecosystems in South Africa have been fenced and divided into smaller man-made systems such as farms, parks and reserves. These artificial measures prevent species from migrating freely and disturbances such as the introduction of unadapted game species and/or stock demand careful management of man-made ecosystems. Any further disturbance such as overgrazing will have a ripple effect, to the detriment of all aspects of the biotic and abiotic components. Effective management of a man-made ecosystem requires a thorough knowledge of especially the interaction within ecosystems.

PLANT SUCCESSION

Plant communities, as other communities in an ecosystem, are dynamic and continually subject to change. Changes may occur when conditions for a specific plant community become unfavourable as a result of disturbances like overgrazing and drought. The disturbed area is then invaded by another, better adapted plant community. As conditions improve, a succession of plant communities takes place progressively until a climax community has been established. This following of plant communities in an area, is known as plant succession. It is especially noticeable in the different grass communities.

The succession process begins when a plant community establishes itself on bare soil, for example bare patches in veld or cultivated land. The first plant community is known as the pioneer community and represents the pioneer stage of plant succession. Plants in the pioneer community usually consist of annual grasses and herbs, such as Annual Three-awn (*Aristida adscensionis*) and Khaki weed (*Tagetes minuta*), adapted to extremely unfavourable conditions. Conditions are improved by the pioneer community, as the vegetation covers the soil and thus protects it against sun and wind. The vegetation cover furthermore decreases run-off of rain water, thereby ensuring a higher infiltration of moisture into the soil. Fertility of the soil is also increased to some extent by the production of organic material. These improved conditions pave the way for establishment of one or more subclimax communities.

Subclimax communities form an intermediate stage between the pioneer and the climax stages and prepare the ideal growing conditions for the climax community.

The climax community represents the final or climax stage of the plant succession process. Grasses of the climax community, for example Rooigras (*Themeda triandra*), are mainly perennial and well adapted to the normal growing and climatic conditions. Any disturbance during plant succession will push the process back towards the pioneer stage.

There is a distinct correlation between the grazing value of an area and the stage of succession of the vegetation. Grasses in the pioneer community usually have a short growth cycle and depend mainly on seed production for survival. Little energy is made available by pioneer plants for leaf production, with a consequently low leaf mass yield and therefore also a low grazing value (also see Chapter 4). Climax species on the other hand, are mainly perennial and depend on leaf production for survival: the leaves produce reserve nutrients which can be used for regrowth in the next growing season. This survival mechanism leads to higher leaf production and therefore also to higher grazing value. A climax community will also limit soil erosion to the minimum and ensure maximum utilization of available moisture. Healthy climax communities are better adapted than subclimax and pioneer communities to survive natural disturbances such as droughts.

In many areas, continuous overgrazing of a plant community will bring about a desert community, in spite of the ability of the local climate to support a grassland community (Ryke, 1978). Such a desert community is known as a disclimax community.

SWEETVELD, SOURVELD AND MIXED VELD

For grazing purposes, the natural veld in South Africa can be divided into three broad veld types, namely sweetveld, sourveld and mixed veld (Figure 13). These veld types differ from each other mainly in the nutritive value and palatability of the common grasses during the unfavourable season.

Sweetveld, sourveld and mixed veld originate primarily through the adaptation of plant species to climatic factors such as rainfall and temperature and the influence of these factors on the environment. As a result of the relatively higher rainfall in the sourveld areas, the soil is constantly subject to the leaching of plant nutrients, with a resultant decrease in the pH and fertility of the soil. Leaching of plant nutrients has a negative influence on the nutritive value and palatability of grasses. Severe winters with frost are typical of sourveld areas. Grasses adapt to this by translocation of nutrients. Reserve nutrients are stored in the roots and leaf bases to survive the unfavourable period. In winter most sourveld grasses therefore have very little nutrients available as grazing.

Only a slight amount of leaching of plant nutrients in the soil and translocation of nutrients in the plant occur in sweetveld, because the rainfall is lower and the temperature usually higher than in sourveld areas. Palatability and nutritive value of grasses in sweetveld are therefore higher than in sourveld. Winter grazing can therefore be applied more successfully in sweetveld than in sourveld.

Sweetveld, sourveld and mixed veld are characterized by the following:

Sweetveld
- Occurs mainly in low-lying and nearly frost-free areas.
- Rainfall ranges from about 250—500 mm per annum.
- Most grasses provide palatable grazing throughout the year, if the veld is in good condition.
- Is sensitive to overgrazing, especially during the growing season.
- Recovers more quickly from disturbance than sourveld, if optimal growing conditions prevail.

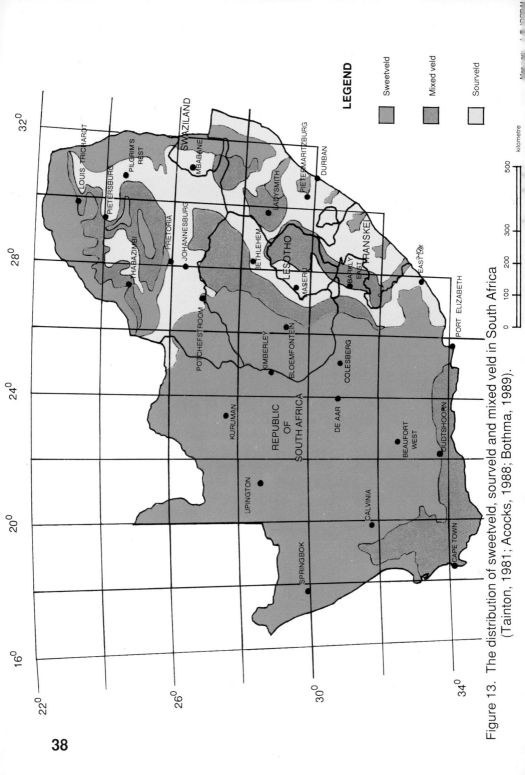

Figure 13. The distribution of sweetveld, sourveld and mixed veld in South Africa (Tainton, 1981; Acocks, 1988; Bothma, 1989).

Sourveld

- Occurs at higher altitudes and at lower temperature than sweetveld.
- Rainfall about 625 mm and more per annum.
- Most grasses produce palatable grazing with a fairly high nutritive value only in the growing season.
- Can tolerate overgrazing, but this will lead to lower production.
- Recovers more slowly from disturbances than sweetveld.

Mixed veld

Mixed veld represents an intermediate form between sweetveld and sourveld. Mixed veld of which the characteristics correspond to those of sweetveld, is known as sweet mixed veld, and vice versa as sour mixed veld.

BIOMES OF SOUTH AFRICA

Biomes are formed by the influence of climate on the living and non-living components of a region, and by the adaptation of especially the plant community to these climatic conditions. A biome is the largest land community and is characterized by the uniformity of the general climax vegetation. In the grassland biome, for example, grasses are characteristic, although the dominant species may vary from area to area. However, a biome does not consist of vegetation only, but includes all living organisms who find a suitable habitat in it. The six different biomes recognized in South Africa are shown in Figure 14 and are briefly discussed (Rutherford & Westfall, 1986).

Forest Biome

The vegetation of the Forest Biome consists mainly of evergreen plants and is characterized by trees forming a continuous canopy of leaves. The Forest Biome is by far the smallest biome in South Africa. Apart from the area in the Southern Cape (Knysna), numerous smaller forest patches are found, especially in the eastern and southern parts of the country. Forests occur from sea-level up to an altitude of 2 134 m. The mean annual rainfall of forest areas in the winter rainfall region is 525 mm or more; in the summer rainfall region it is 725 mm or more (Figure 15). Frost is absent in the Forest Biome.

Fynbos Biome

The term fynbos is used for vegetation with fine leaves. The Fynbos Biome is usually confined to mountainous areas and is characteristic of the south-western and southern Cape Province. Fynbos occurs from sea-level up to an

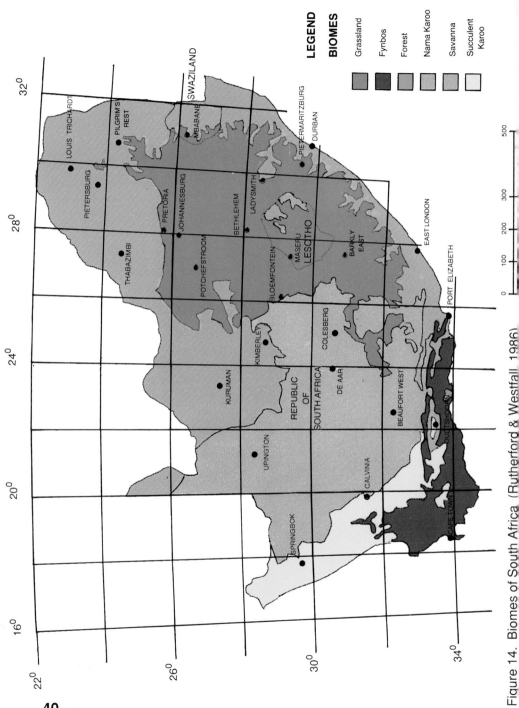

Figure 14. Biomes of South Africa (Rutherford & Westfall, 1986)

altitude of 2 325 m. The mean annual rainfall ranges from 210—3 000 mm and rain occurs mainly in winter. Frost is rare in the coastal areas, but light frost is often experienced in the higher parts. The carrying capacity of veld in the Fynbos Biome is very low. Most plant species are unpalatable and unacceptable as grazing and have to supplemented by artificial pasture. More than half the plants that occur in the Fynbos Biome are unique to this biome.

Grassland Biome

The term grassland refers to veld in which the dominant plant species are grasses. Although the Grassland Biome occurs at a minimum altitude of 300 m in the eastern parts of South Africa, it is particularly characteristic of the central plateau with an altitude of up to 2 850 m. This biome is confined to the summer rainfall parts with an mean annual precipitation of 400—2 000 mm. Frost commonly occurs in winter. Although most grasses in this biome are well adapted to grazing, injudicious veld management will have a detrimental effect on the species composition.

Nama Karoo Biome

Nama Karoo is a concatenation of Namaland in Namibia and the Karoo of South Africa. The vegetation in this biome can be seen as grass shrub veld. It occurs on the central plateau of the Cape Province, the south-western part of the Orange Free State and the southern interior of Namibia. The altitude ranges from 1 000 to 1 400 m. Part of the Nama Karoo Biome also occurs in Lesotho at 2 850—3 400 m above sea-level. Rain occurs either throughout the year or in summer, with an annual mean of 100—520 mm. Minimum temperatures are low and frost is regularly experienced in winter. The largest part of the biome is used for grazing, mainly for sheep and goats.

Savanna Biome

The term savanna (bushveld) describes veld of which the vegetation comprises a herbaceous layer (mostly grasses) and a woody layer (mostly trees). The canopy cover of the woody layer may vary from widely spaced to a cover of 75%. In South Africa the Savanna Biome extend over the higher-rainfall parts of the Northern Cape, the northern two thirds of Transvaal and the low-lying parts of Natal and the North-western Free State. The altitude ranges from sea-level to 2 000 m. The largest part of the Savanna Biome occurs in summer rainfall areas with a mean annual precipitation of 235 mm or more. The Savanna Biome is absent from areas where high rainfall and low winter temperatures are experienced, but does occur in parts with low rainfall and severe frost. Most of the Savanna Biome is used for extensive utilization of vegetation by livestock and game.

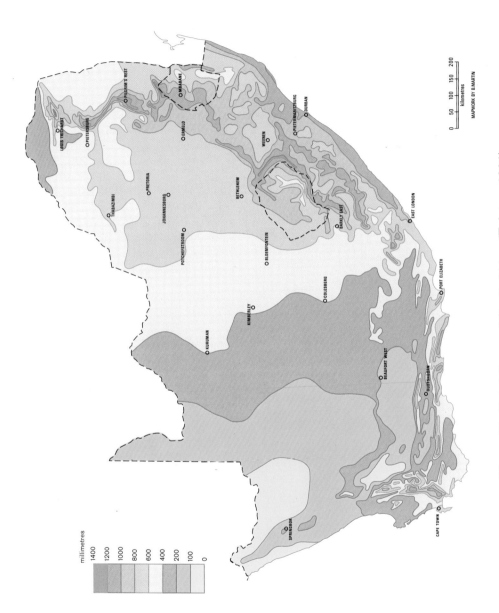

Figure 15. Rainfall map of South Africa (Tainton, 1981).

Succulent Karoo Biome

The Succulent Karoo Biome or 'Winter Rainfall Karoo' is characterized by the numerous succulent plants and the low occurence of perennial grasses. Most of this biome lies to the west of the western escarpment and the landscape is generally flat to undulating. The altitude ranges from sea-level to 1 500 m, but most of the biome lies below 800 m. The mean annual rainfall ranges from 20—290 mm and it occurs mainly in winter. The veld is utilized primarily by sheep and goats, but the grazing capacity is very low. The species diversity of succulent plants in this biome is among the highest in the world.

OPTIMAL GRAZING

Optimal grazing is the utilization of herbaceous vegetation (usually grasses) without the deterioration of the productivity and vigour of the plants utilized. Optimal grazing can be applied only if veld management practices are adapted to the grazing habits of game species as these occurred before the arrival of man. The pasture in South Africa has been subjected to the grazing habits of game species for thousands of years and is therefore well adapted to these habits. However, information about this is scant and one often has to speculate or make one's own deductions.

Because the grazing habits of nearly all game species differ from one another to a greater or lesser extent, the veld could support a wide variety of game, without direct competition for food between them. In the past, the veld could therefore support more game than livestock (in large stock units) without adverse effects. Furthermore, many game species limit deterioration of the veld by driving surplus animals out of the herd or territory. These surplus animals are often old or sick and therefore easy prey for predators. Since the availability of surface water was often limited in the early days, especially at certain times of the year, some water-dependent species had to migrate from one water point to another. Most game species also habitually move to areas that offer better grazing; consequently overgrazing was rare in those early times.

It can be inferred that in the past the veld had a relatively low and balanced occupancy of game, but at times was subjected to periods of relatively high occupancy of migrating grazers, followed by long periods of veld rest. Since the veld is well adapted to the above-mentioned grazing methods, any deviation such as overgrazing without allowing the veld to rest, will have a negative influence on especially the productivity of the grassland. In the long term, such deviations usually cause highly unproductive side-effects such as bush encroachment, bare soil and/or soil erosion.

CARRYING CAPACITY

Carrying capacity of veld refers the potential of an area to support livestock and/or game through the utilization of vegetation, without deterioration of the ecosystem in the long term. The potential carrying capacity of an area is expressed in large stock unit per hectare (AU/ha) or hectare per large stock unit (ha/AU).

The carrying capacity of an area is determined by the annual rainfall and the composition of usable plant species occurring in the area. Since the composition of usable species differs from farm to farm and even from one camp to the next, a carrying capacity has to be determined for every individual farming unit. The carrying capacity of an area also changes annually as the annual rainfall varies, and therefore stocking rates have to be adjusted constantly.

REFERENCES AND LITERATURE CONSULTED

ACOCKS, J.P.H. 1988. Veld types of South Africa. **Memoirs of the Botanical Survey of South Africa** no. 57
BOTHMA, J. DU P. 1989. **Game ranch management.** J L van Schaik, Pretoria
ODUM, E.P. 1971. **Fundamentals of ecology.** Saunders, London.
RUTHERFORD, M.C. & WESTFALL, R.H. 1986. Biomes of Southern Africa: an objective categorisation. **Memoirs of the Botanical Survey of South Africa.** no. 54
RYKE, P.A.J. 1978. **Ekologie - beginsels en toepassings** Butterworths, Durban
TAINTON, N.M. 1981. **Veld and pasture management in South Africa.** Shuter & Shooter, in co-operation with University of Natal Press, Pietermaritzburg.
TROLLOPE, W.S.W., TROLLOPE, LYNNE & BOSCH, O.J.H. 1990. Veld and pasture terminology in Southern Africa. **J. Grassl. Soc. South Africa** 7(1): 67—75

3 VELD MANAGEMENT IN GRASSLAND AND SAVANNA AREAS

Veld management refers to the management of natural vegetation for specific objectives related to different forms of land use (Trollope *et al.*, 1990). In the agricultural context the objective of veld management is the sustained production of livestock products through the maintenance of the veld in a stable and productive condition. In the wildlife context a wide spectrum of different forms of land use are practised, ranging from pure nature conservation to the game ranching. Consequently the objectives of veld management vary greatly, from creating and maintaining plant communities attractive to tourists, trophy hunting and the production of venison and other animal products. Nevertheless, correct veld management is the key to successful achievement of these objectives and a thorough understanding of this topic is an essential prerequisite for this vigorous and stable ecosystem. In this chapter emphasis will be given to the management of the grass sward for livestock production and game ranching in the grassland and savanna areas of South Africa, because these constitute the major forms of land use, and grasses comprise the dominant source of forage produced in these areas.

Veld management comprises three components and these are used to achieve the previously mentioned objectives. The components are veld management practices, systems and layouts. The two latter components generally apply only to livestock production systems, while the former is an important element in both livestock and wildlife systems of land use. Veld management practices are the treatments that are applied in the management of veld and comprise stocking rate, animal ratios, rotational grazing, rotational resting and veld burning. Veld management systems are the formalized programmes through which veld management practices are applied and refer to the number of camps allocated per group of animals. Veld management layouts are the arrangement of camps in a veld management system in the field.

The initial step in the formulation of a veld management programme, is the assessment of the condition of the veld in each homogeneous vegetation unit of the area to be planned.

VELD CONDITION

Veld condition refers to the condition of the vegetation in relation to some functional characteristic (Trollope *et al.*, 1990). In livestock production systems the two most important functional characteristics are forage production and resistance to soil erosion. These also apply to wildlife management, but in this case two additional functional characteristics are recognised, namely the physiognomic structure of the vegetation and the potential of the veld to produce grass fuel to support a fire. The physiognomic structure of the grass sward indicates the potential of the vegetation to provide suitable habitat for different wildlife species. The potential to produce grass fuel indicates the ability of the vegetation to support a fire, for example in the control of bush encroachment.

One of the most widely used techniques for assessing veld condition in terms of its potential to produce grass forage, grass fuel and resist soil erosion, is that developed by Foran *et al.* (1978) and adapted for use in the Eastern Cape, Ciskei and Kruger National Park by Trollope (1986), Danckwerts (1987) and Trollope *et al.* (1989).

In the agricultural sphere, the assessment of the condition of the herbaceous layer is based on the botanical composition and basal cover of the grass sward. It involves comparing the condition of the veld in a **sample site** with that of a **benchmark site** which represents the potential condition of the grass sward for optimum livestock production in a particular veld type. Emphasis is given to determining the botanical composition of the grass sward, because Danckwerts (1981) found that this parameter is a good indicator of the inherent ability of the veld to produce forage for grazing ungulates. The basal cover serves as an index of the resistance of the veld to soil erosion, and is estimated subjectively.

A similar approach has been used for assessing veld in wildlife situations, but in this case the condition of the sample sites is not compared with that of a benchmark site, and the forage and fuel production potentials of the veld are rather indicated by forage and fuel scores (Trollope *et al.*, 1989). A further refinement to the techniques has been the identification of key grass species which have the greatest effect on veld condition. These limited number of species are used to assess the condition of the veld and have resulted in the simplification of the techniques for easier use by land users. In both cases the techniques also provide a measure of the ecological status of the veld in terms of whether it is or has been correctly, heavily, lightly or selectively grazed. This is achieved by categorizing the grasses into so-called **decreaser** and **increaser** species. Decreasers are grass species which decrease when the veld is under- or overgrazed. Increaser I's are grass species which increase with under- or selective grazing, and Increaser II's are grass species which increase with overgrazing (also see Chapter 4, ecological status).

A good indicator of the physiognomic structure of the grass sward is the standing crop of grass, as it describes the volume and density of plant material at ground level. Results obtained by De Wet (1988) in the Kruger National Park, showed that the standing crop of grass is closely related to the habitat preferences of game species like buffalo and wildebeest which prefer tall and short grass respectively. A rapid technique for estimating the standing crop of grass in the veld, is the disc pasture meter developed by Bransby & Tainton (1977). This apparatus relates the settling height of an aluminium disc to the standing crop of grass holding it above the ground. A standard calibration for the disc pasture meter has been developed for the Kruger National Park (Trollope & Potgieter, 1986), and experience has shown that this calibration yields satisfactory results elsewhere in the Transvaal bushveld. A standard calibration has also been developed for use in the Eastern Cape (Trollope, 1984).

These veld condition data are a prerequisite for the planning of a sound veld management programme and are used in the formulation of veld management practices.

VELD MANAGEMENT PRACTICES

Stocking rate
Stocking rate refers to the area of land in the system of management which the operator has allocated to each animal unit in the system (Booysen, 1967). The stocking rate of grazing species on a livestock or game ranch is dependent on the grazing capacity of the veld which is in turn influenced by the condition of the grass sward. This is not necessarily true for low density game species; in this case optimal habitat conditions are the most important requirement for the successful establishment and maintenance of a game population (Bothma, 1989). Nevertheless, veld condition is generally of paramount importance in determining the stocking rate of grazers and must be based on the grazing capacity of the veld.

Considerable progress has been made in the agricultural field in the estimation of grazing capacity based on veld condition; the procedures are fully described by Trollope (1986) and Danckwerts & Teague (1989). It is believed that with appropriate adaptations, the same procedure could be used for estimating the grazing capacity of veld for wild ungulates and for the formulation of realistic stocking rates for game populations.

Animal ratio
Animal ratio is the ratio of bulk grazers to concentrate grazers (Table 2). The selective grazing habits of domestic and wild ungulates are related to the grazing sequence in which these different classes of animals utilise a pasture. It normally involves heavy bulk grazers preceding light concentrate grazers, thus preparing the pasture for use by those animals that follow. Research in the Eastern Cape showed that in the sourveld areas the ratio of cattle to sheep

should not exceed 1 Large Stock Unit (AU) : 6 Small Stock Units (SSU) (Danckwerts & Teague, 1989), while experience in the sweetveld areas indicates that it should not exceed 1 AU : 3 SSU, because in arid areas veld is more susceptible to selective grazing. In the wildlife field, Mentis (1981) recommended that the metabolic mass of concentrate grazers should not be permitted to exceed that of bulk grazers in any grazing unit. On the basis of research results from the Eastern Cape, he proposed a maximum ratio of 1 AU bulk grazers : 1 AU concentrate grazers. However, in view of earlier discussion, it is recommended that in wildlife areas a maximum ratio of 1 AU bulk grazers : 1 AU concentrate grazers be used in sourveld areas, and a maximum ratio of 1 AU bulk grazers : 0,5 AU concentrate grazers in sweetveld areas.

Table 2. Classification of some ungulates according to their grazing habits (Mentis, 1981).

Concentrate grazers	Bulk grazers
Blesbok	Cattle
Blue wildebeest	Burchell's zebra
Bushpig	Buffalo
Oryx	Waterbuck
Nyala	
Oribi	
Reedbuck	
Impala	
Mountain reedbuck	
Sheep	
Springbuck	
Black wildebeest	
Grey rhebok	
Warthog	
White rhinoceros	

Rotational grazing

Rotational grazing is the successive occupation of different areas by a group of animals during the year so that not all the veld is grazed simultaneously (Tainton, 1981). The basic objective is the improvement and/or maintenance of the condition of the veld, and generally this form of management is applied only in livestock production systems because it is difficult to apply in wildlife situations. The are two forms of rotational grazing, namely:

High production grazing (HPG)- this is the occupation of a camp by grazing animals until all the acceptable and desirable grass species have been grazed to a stage that will ensure rapid regrowth and high production of forage (Tainton, 1981).

High utilization grazing (HUG)- this is the occupation of a camp by grazing animals until all the grass species have been heavily grazed (Tainton, 1981).

The use of different forms of rotational grazing depends upon the condition of the veld and specifically the dominance of the different categories of grass species, i.e. decreaser and increaser species. Of course, besides the application of the correct form of rotational grazing, appropriate adjustments must also be made to the stocking rate in accordance with the dominance of decreaser and increaser grass species. The use of different forms of rotational grazing under different veld conditions is illustrated in Figure 16.

When the veld is dominated by decreaser species, a combination of high production grazing and high utilization grazing must be applied. High production grazing is used for the majority of the time to maximize forage pro-

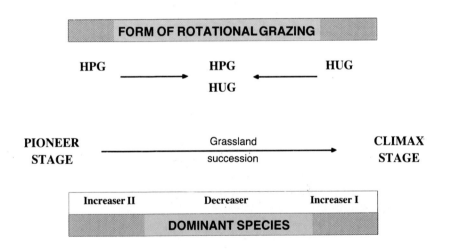

Figure 16. Different forms of rotational grazing to be applied when veld is in different stages of the grassland succession.

duction and animal performance, while high utilization grazing is applied occasionally to stabilise the grassland succession at the decreaser stage.

When the veld is dominated by Increaser I species, high utilization grazing must be applied to reverse the grassland succession back to the decreaser stage. This can be combined with veld burning or the feeding of nitrogen/protein licks so as to minimize the negative aspects of this form of rotational grazing on animal performance.

Finally, when the veld is dominated by Increaser II species, high production grazing must be applied to encourage the development of decreaser species and move the succession to the decreaser stage.

Rotational resting

Rotational resting is the successive withdrawal of veld from grazing on a rotational basis for specific purposes. The principal reasons for resting veld are seeding, restoration of plant vigour and provision of a fodder reserve for periods of scarcity. Rotational resting is an important and essential veld management practice in both livestock and wildlife management systems.

A seeding rest is a priority when the decreaser species have declined in the grass sward and is appropriate when the veld is dominated by either Increaser I or Increaser II grass species.

Resting the veld to restore plant vigour, is necessary under all veld conditions and is aimed at the decreaser species. In the sweetveld areas this necessitates resting the veld for a full year so that the grass plants can grow to maturity at which stage they are still valuable forage for livestock. In the mixed and sourveld areas, a shorter rest during the late summer and autumn period is adequate for this purpose and reduces the extent to which the grass becomes mature and unacceptable to livestock.

A fodder reserve rest is necessary under all veld conditions, but enjoys priority when the veld is in excellent condition and seed production is not important. It is intended to provide a fodder reserve for that time of the year when there is the greatest likelihood of a deficiency.

The implementation of a rotational resting programme is more easily applied in a livestock production system where the grazing area has been subdivided into grazing camps. Under these circumstances experience has shown that one third of the grazing area should be rested annually in sweetveld areas and one quarter in sourveld areas, according to the previously discussed guidelines. The greater resting intensity in sweetveld areas is aimed at minimising the negative effects of the lower and more erratic rainfall occurring in these areas.

In wildlife areas the application of rotational resting is more difficult: the veld has to be treated on a rotational basis in such a way that game will be attracted to these areas, thus providing a rest period for the vacated area. Generally the most effective way of attracting game to a particular area is to burn the veld, resulting in the production of highly palatable and nutritious forage. Care must be taken to ensure that the burnt area exceeds the short term forage requirements of the game that will be attracted to it, so that no overgrazing of the veld occurs. Rotational resting can also be implemented by the strategic placing of licks to attract game to a new area, and by the opening and closing of watering points (Young, 1989).

Veld burning

Veld burning is an important and often essential veld management practice in both livestock and wildlife management systems. The most important factors to consider when developing a burning programme, are the seasons for burning and the appropriate fire regime to be applied. There are basically two reasons for burning veld (Trollope, 1989):
- to remove moribund and/or unacceptable grass material, and
- to eradicate and/or prevent the encroachment of undesirable plants.

The fire regime to be used, refers to the type and intensity of fire and the reason and frequency of burning.

Type of fire. In grassland and savanna areas, surface fires are the most common type of fire. In this category there are head fires which burn with the wind, and back fires which burn against the wind. It is recommended that head fires be used in all cases, because they cause least damage to the grass sward but can cause maximum damage to woody vegetation if required (Trollope, 1989).

Fire intensity. Fire intensity refers to the rate at which heat energy is released per unit length of fire front and is expressed in kilojoules per second per metre (kJ/s/m). When burning to remove moribund and/or unacceptable grass material, a cool or low intensity fire of less than 1 000 kJ/s/m is recommended. This can be achieved by burning when the air temperature is lower than 20 °C and the relative humidity higher than 50%. When burning to control undesirable plants like encroaching bush, a high intensity fire of more than 2 000 kJ/s/m is required. This can be achieved when the grass fuel load is more than 4 000 kg/ha, the air temperature is 25—30 °C and the relative humidity less than 30% (Trollope, 1989). This will cause a significant topkill of stems and branches of bush species up to a height of 3 m. In all cases the wind speed should not exceed 20 km/h.

Season of burning. Research in Natal and the Eastern Cape has shown that the least damage is caused to the grass sward if burning is applied when the grass is dormant. Therefore it is recommended that burning to remove moribund and/or unacceptable grass material should preferably be applied after the first spring rains when the grass is still dormant. Conversely, burning to control encroaching plants should be applied before the first spring rains while the grass is very dry and dormant, to ensure a high intensity fire (Trollope, 1989).

Frequency of burning. When the aim is to remove moribund and/or unacceptable grass material, the frequency of burning will depend upon the accumulation rate of excess grass litter (Trollope, 1989). Field experience indicates that this should not exceed 4 000 kg/ha, and therefore the frequency of burning should be based on the rate at which this phytomass of grass material accumulates. This approach has the advantage that the frequency of burning is related to the stocking rate of grazers and to the amount of rainfall the area receives. Generally in sourveld areas this will result in a frequency of burning of every 2—4 years, but in sweetveld areas it will be much lower. This rule of thumb will exclude fire where the condition of the veld is so poor that excessive grass fuel loads never accumulate. Therefore, in terms of rotational resting, the frequency of burning will be one of the determining factors in the application of this veld management practise in wildlife areas.

Post-fire grazing management. Post-fire grazing management probably has a greater effect on the condition of the grass sward than any other aspect of veld burning. Despite this, very little quantitative information is available on the effects of different grazing and burning regimes on the grass sward. Preliminary results obtained in the sourveld areas of the Eastern Cape suggest that continuous grazing after burning encouraged Increaser II grass species (*Eragrostis plana*) at the expense of decreaser species (*Themeda triandra*). Conversely, rotational grazing comprising high utilization grazing for short periods of time, resulted in an increase of decreaser species (*Themeda triandra*). Furthermore, commencing rotational grazing after burning when the grass was short (5 cm), resulted in less selective grazing than grazing when the grass was longer (10—15 cm) (Danckwerts pers. comm. 1990). It is therefore recommended for livestock production systems that, when burning to remove moribund and/or unacceptable grass material, rotational grazing for short periods of time (one week) be applied soon after the burn, allowing adequate recovery between grazings. However, when burning to control undesirable plants, post-fire grazing management will depend upon the ecological characteristics of the encroaching plant.

Post-fire grazing management in wildlife areas is difficult to control. To prevent overgrazing, it is important to ensure that the burnt area exceeds the short term forage requirements of the game that will be attracted to it.

Monitoring veld condition

The recent development of techniques for assessing veld condition is one of the most important advances in the field of veld management. This is because veld condition data are important not only in the planning of a veld management programme, but trends in veld condition monitored over time can also be used to evaluate and adapt veld management practices when necessary. It is therefore essential that the condition of the veld be monitored on a regular basis, both in agricultural and wildlife areas.

VELD MANAGEMENT SYSTEMS

Veld management systems are generally relevant only to livestock production systems and refer to the number of camps allocated per group of animals. There are two broad categories, namely pauci-camp and multi-camp systems. Pauci-camp systems are systems with fewer than five camps per group of animals, with three-camp one-herd systems recommended in sweetveld areas, and four-camp one-herd systems in sourveld areas. Multi-camp systems are systems with more than five camps per group of animals, with twelve-camp one-herd systems recommended in sweetveld areas, and eight-camp one-herd systems in sourveld areas. Generally multi-camp systems are preferred to pauci-camp systems, because selective grazing can be more easily controlled and they provide greater managerial flexibility. However, multi-camp systems are more expensive in terms of fencing and the provision of watering points. In conclusion it must be stressed that the veld management practices involved in the application of a veld management system are more important than the system itself.

VELD MANAGEMENT LAYOUTS

Veld management layouts are also generally relevant only to livestock production systems, and the arrangement of camps can be divided into two broad categories, namely conventional and 'wagon wheel' layouts (Figure 17). The conventional layout of camps involves subdivision of the veld into homogenous vegetation units which serve as the basic units for development of pauci- or multi-camp systems. Normally each camp is provided with its own livestock drinking point, or it may be shared between camps depending on circumstances. The 'wagon wheel' layout of camps generally involves a multi-camp system with a central hub from which camps radiate to the perimeter of the grazing area. A single drinking point is located in the central hub, and other livestock handling facilities can also be developed in this area for ease of management.

The greatest advantage of the 'wagon wheel' layout is the ease with which the veld and the livestock can be managed in terms of inspecting the state of the grazing in the camps and the easy movement and handling of livestock. Having only one drinking point, is also a great advantage in terms of development and maintenance costs and there is a significant saving in labour because of the efficient mustering and handling of livestock. The greatest disadvantage of the 'wagon wheel' layout is the greater potential for area selective grazing, because it is more difficult to separate different types of veld into individual camps. This type of layout is therefore better suited to flat areas than to undulating and broken terrain. This problem can be minimized with careful and imaginative planning. Another serious disadvantage of the 'wagon wheel' layout is that it requires more fencing.

It is difficult to conclude as to which is the more desirable layout; suffice it to say that when planning a veld management programme, both types of layout should be considered and the final choice made on the basis of the particular circumstances prevailing in the area in question.

VELD REHABILITATION

Veld rehabilitation is the restoration of degraded veld to a productive and stable condition (Trollope *et al.*, 1990). This is achieved through the development of a dense, perennial, vegetal cover comprising highly acceptable, nutritious and productive plants. In practice this generally consists of controlling the encroachment of undesirable plants and the use of mechanical techniques to rehabilitate veld are only resorted to when biological methods have failed. The most important factor that influences the adoption and implementation of a veld rehabilitation programme, is economics. Unless veld rehabilitation techniques

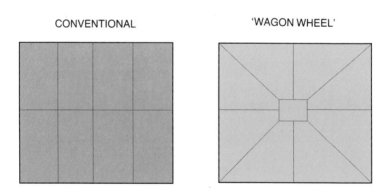

Figure 17. Diagrammatic representation of a conventional and 'wagon wheel' layout of an eight-camp one-herd system.

are compatible with the economic capabilities and aspirations of the land user, there is little chance that they will be implemented. An assessment of the economic characteristics of livestock production off veld and the economic returns from wildlife management systems, shows that the potential and actual return per unit area (R/ha) of veld is low. This economic reality demands that the technology used in veld rehabilitation of necessity be non-capital intensive. In most cases the only economically feasible methods for rehabilitating veld on an extensive scale, are the implementation of appropriate veld management practices, i.e. stocking rate, rotational grazing and resting, stocking with appropriate ungulate species, and veld burning. The actual form that these practices take, depends on the condition of the veld. This emphasises the importance and necessity of assessing the condition of the veld and developing a monitoring programme for evaluating veld rehabilitation programmes, and adapting them in the light of such a monitoring programme.

REFERENCES

BOOYSEN, P. DE V. 1967. Grazing and grazing management terminology in South Africa. **Proc. Grassl. Soc. South Africa** 2: 45—57.
BOTHMA, J. DU P. (ed.) 1989. **Game ranch management**. Van Schaik, Pretoria.
BRANSBY, D.I. & TAINTON, N.M. 1977. The disc pasture meter: possible applications in grazing management. **Proc. Grassl. Soc. South Africa** 12: 115—118.
DANCKWERTS, J.E. 1981. **A technique to assess the grazing capacity of sweetveld with particular reference to the false thornveld areas of the Ciskei.** M.Sc.(Agric.) thesis, University of Natal, Pietermaritzburg.
DANCKWERTS, J.E. 1987. A revised method for assessing the condition of the herbacous vegetation in the grassveld and savanna areas of the Eastern Cape Region. **Dohne Agric.** 9(2): 10—15.
DANCKWERTS, J.E. & TEAGUE, W.R. 1989. **Veld management in the Eastern Cape.** Government Printer, Pretoria.
DE WET, S.F. 1988. **Habitat en voedingsvoorkeure van die blouwildebees (Connochaetes Taurinus Taurinus Burchell, 1923) en ander diersoorte in die sentraal suidoostelike gebied van die Nasionale Krugerwildtuin.** M.Sc. thesis, University of Pretoria, Pretoria.
FORAN, D.B., TAINTON, N.M. & BOOYSEN, P. DE V. 1978. The development of a method for assessing veld condition in three grassveld types in Natal. **Proc. Grassl. Soc. South Africa** 13: 27—33.
MENTIS, M.T. 1981. The animal as a factor in pasture and veld management. In N.M. Tainton (ed.), **Veld and pasture management in South Africa**. Shuter & Shooter, Pietermaritzburg.
TAINTON, N.M. 1981. **Veld and pasture management in South Africa**. Shuter & Shooter, Pietermaritzburg.

TROLLOPE, W.S.W. 1984. **Control of bush encroachment with fire in the arid savannas of south-eastern Africa**. Ph.D. thesis, University of Natal, Pietermaritzburg.

TROLLOPE, W.S.W. 1986. Land use surveys: assessment of veld condition in Ciskei. In **Republic of Ciskei National Soil Conservation Strategy**, Vol. 1. Dept. of Agriculture & Forestry, Ciskei.

TROLLOPE, W.S.W. 1989. Veld burning as a management practice in livestock production. In J.E. Danckwerts, J.E. & W.R. Teague (eds.), **Veld management in the Eastern Cape**. Government Printer, Pretoria.

TROLLOPE, W.S.W. & POTGIETER, A.L.F. 1986. Estimating grass fuel loads with a disc pasture meter in the Kruger National Park. **J. Grassl. Soc. South Africa** 3(4): 148—152.

TROLLOPE, W.S.W., POTGIETER, A.L.F. & ZAMBATIS, N. 1989. Assessing veld condition in the Kruger National Park using key grass species. **Koedoe** 32(1): 67—93.

TROLLOPE, W.S.W., TROLLOPE, LYNNE & BOSCH, O.J.H. 1990. Veld and pasture terminology in southern Africa. **J. Grassl. Soc. South Africa** 7(1): 52—61.

YOUNG, E. 1989. **Wildboerdery en natuurreservaatbestuur**. Eddie Young Uitgewers, Nylstroom.

4 GRAZING VALUE AND ECOLOGICAL STATUS OF GRASSES

Grasses are generally well adapted to optimal grazing and other methods of defoliation. The growing zones of grass plants are situated at or even below the surface of the soil, and are therefore protected from grazers and fire. Although grasses are adapted to optimal grazing, the greatest threats to the grass plant remain undergrazing and especially overgrazing.

When no grazing or other methods of defoliation take place, it leads to the accumulation of organic material which smothers the grass plant, and it eventually dies. During overgrazing, regrowth of the plant is repeatedly overutilized, and reserve nutrients used for regrowth therefore become depleted. This weakens the root system and the plant eventually perishes. Although many grasses are not adapted to overgrazing, there are a few grass species that delay grazing by employing deterrents. Examples of these are unpalatable chemical substances (species of *Bothriochloa*), tough stringy leaves (*Aristida junciformis* subsp. *junciformis*) and leaves with a high breaking stress (*Eragrostis plana*).

There is a clear correlation between the degree of adaption to grazing, the value for grazing (grazing value) and the reaction to grazing (ecological status). For example, most grass species well adapted to overgrazing, will have a low grazing value and will increase during overgrazing.

The grazing value and the ecological status of a grass species are greatly subjective and difficult to determine, because of the numerous factors involved. However, this is information that may hold important management implications for the land user. By taking into account the grazing value and ecological status of dominant grasses in grassland and savanna areas, more realistic stocking rates can be established. For the purpose of this book, a grazing value and an ecological status have therefore been allocated to as many grass species as possible. This has been done by consulting grass experts throughout the country. However, these are merely general guidelines, not scientific data, and therefore do not provide for exceptions.

GRAZING VALUE

What is grazing value?

Grazing value can be defined as the potential genetic ability of a grass species to produce grazeable plant material. It can also be seen as the value of a grass species for grazing by animals, in this case especially bulk-grazers such as cattle, buffalo and zebra. The grazing value in this context, is a non-seasonal, long term value under normal, natural growing conditions.

For practical purposes, the grazing value of grass species can be divided into the following six classes:
- very low
- low
- average
- high
- very high
- varying

Factors that determine grazing value

The following genetic factors play a decisive role in the grazing value of a grass species:

Leaf production ability as genetic characteristic, is the ability to produce, under specific conditions, few or many leaves in comparison with other grass species under similar conditions.

Palatability is the general acceptability of a grass species for grazing by animals; it is influenced by factors such as:
- nutritive value
- fibre content
- unpalatable chemical substances
- moisture content

Nutritive value as a genetic characteristic, is determined mainly by the ability of a grass plant to absorb nutrients from the soil, to produce it through photosynthesis and then to make these available to grazers.

Growth vigour is the ability of a grass species to recover quickly after it has been grazed or burnt, and to be available as grazing once again.

Digestibility is influenced mainly by the fibre content of the leaves and culms; the higher the fibre content, the lower the digestibility and vice versa.

Perenniality. Generally, more leaf material is produced by perennial grasses than by annual grasses, the latter being directed more towards seed production than towards leaf production.

Habitat preference. The habitat to which a species is adapted will have a decisive influence on its grazing value, since the habitat affects the above-mentioned factors such as leaf production ability, nutritive value and palatability.

Variation in grazing value

The grazing value of most grasses is constant, because they vary little genetically and are regional or habitat specific. Nevertheless, there are a number of grasses of which the grazing value varies to a greater or lesser extent. This particularly happens in species with a wide geographical distribution and adapted to a variety of environmental conditions. Variations in the grazing value of a certain species, can be attributed mainly to:

Differences in environmental conditions. Differences in rainfall, temperature, soil, geology, aspect and humidity affect the grazing value of a species.

Genetic variations. The different subspecies, variations, strains, ecotypes and forms of a species may differ in grazing value.

Veld management practices. Bad veld management practices like overgrazing will reduce the growth vigour and leaf production, and therefore also the grazing value of a grass (especially palatable species). Good veld management practices have the opposite effect.

Genetic variation and differences in environmental conditions go hand in hand, since it is specifically adaptation (natural selection) to various environmental conditions that results in genetic variation. This is why a species with a wide geographical distribution, adapted to a variety of environmental conditions by means of genetic variation, can vary in grazing value.

In this guide, the grazing value of species of which the grazing value varies, are for example given as:
- grazing value average to high
- grazing value mostly low
- grazing value varying.

Although grazing value is a long term value, there can be short term variations in the grazing value of a grass species in a specific area. These short term variations are caused by the effect of annual fluctuations in rainfall on the productivity of the grass plant. This tendency particularly applies to the drier parts of the country.

ECOLOGICAL STATUS

A method generally used to determine veld condition, is the 'benchmark site method' (see Chapter 3). This is the comparison of a specific unit of veld or camp (sample site) with a benchmark site that is in optimum condition and occurs in the same ecological zone (Tainton, 1981). By using, among others, type of soil, basal cover, the extent of erosion and the ecological status of dominant grass species, the present veld condition and changes occurring in it over a specific period, can be determined and monitored.

What is ecological status?

Ecological status includes the classification of grasses and forbs into groups on the basis of their reaction to grazing. Grasses generally react to grazing in two ways: a species can decrease or increase in abundance. According to this criterion, all grasses and forbs can be classified into one of the following groups (Trollope *et al.* 1990):

Decreaser. A species that dominates in good veld, but decreases when veld is mismanaged.
Increaser I. A species that dominates in poor veld and increases with understocking or selective grazing.
Increase Ia. A species that increases with moderate understocking or selective grazing.
Increaser Ib. A species that increases with severe understocking or selective grazing.
Increaser II. A species that dominates in poor veld and increases with overgrazing.
Increaser IIa. A species that increases with light overgrazing.
Increaser IIb. A species that increases with moderate overgrazing.
Increaser IIc. A species that increases with severe overgrazing.
Invader. A species that is not indigenous to a specific area.

The grasses in this guide are classified as follows:
- Decreasers
- Increasers I
- Increasers IIb
- Increasers IIc
- Varying

Variation in ecological status

The ecological status of many species is constant. But, as in the case of grazing value, there are some of which the ecological status can vary. These variations occur between regions, veld types and even habitat types. A varying ecological status is therefore especially characteristic of species with a wide geographical distribution and adapted to a variety of environmental and habitat conditions.

Differences in the ecological status within one grass species can be attributed mainly to two factors, namely:

Differences in environmental and habitat conditions. The effect of rainfall, temperature, soil, geology, topography and humidity on the ecological status of a species.

Genetic variation. Different reactions of subspecies, variations, strains, forms and ecotypes to grazing.

These two factors go hand in hand, since adaptation of species to different environmental conditions (natural selection), leads to genetic variation.

In this guide, species of which the ecological status varies, are given as follows, for example:
- Ecological status: Increaser I and IIb.
- Ecological status: mostly Increaser I.
- Ecological status: varying.

It is a well known fact that natural pasture in good condition is the cheapest source of forage for stock and game. For maximum low-cost meat production, it is therefore important for the farmer to provide grasses with a high grazing value to stock and/or game in the long term. This can be done only by improving the veld condition, i.e. the productivity and quality of the veld, by means of a well planned veld management programme (see Chapter 3). An improved or good veld condition will also improve the resistance of an area to soil erosion (see Chapter 1).

REFERENCES AND LITERATURE CONSULTED

DANCKWERTS, J.E. 1989. **Weiding, 'n strategie vir die toekoms: die plant/dier wisselwerking.** Department of Agriculture and Water Supply, Pretoria.

FOURIE, J.H. & VISAGIE, A.F.J. 1985. Weidingswaarde en ekologiese status van grasse en karoobossies in die Vrystaatstreek. **Glen Agric.** 14(1 & 2): 14—18.

JANSE VAN RENSBURG, FRANCI P. & BOSCH, A.J.H. 1989. Influence of habitat differences on the ecological grouping of grass species on a grazing gradient. **J. Grassl. Soc. South Afr.** 7(1): 11—15

TAINTON, N.M. (ed.) 1981. **Veld and pasture management in South Africa.** Shuter & Shooter, Pietermaritzburg.

TROLLOPE, W.S.W., TROLLOPE, LYNNE & BOSCH, O.J.H. 1990. Veld and pasture management terminology in southern Africa. **J Grassl. Soc. South Afr.** 7(1): 52—61.

5 MORPHOLOGY OF THE GRASS PLANT

There is a greater or lesser degree of morphological resemblance between all grasses, making it difficult to distinguish grasses from one another, especially on the species level. The general structure of a grass plant is shown in Figure 18. For the identification of grass species, one often has to use small morphological differences. Morphologically, the grass plant consists of vegetative and reproductive parts.

VEGETATIVE PARTS

Roots

Roots are borne mainly underground and are characterized by the absence of nodes, internodes or leaves. Grasses have an adventitious root system formed by numerous roots developing directly from the culm bases. Depending on the effective depth of the soil, the root system can develop up to 2 metres deep. It's functions are to anchor the plant, absorb moisture and nutrients, and to store reserve nutrients. The following differences with regard to the root system can be used to identify a grass species:

- root development at lower nodes in species with geniculate culms, for example in *Eragrostis lehmanniana* var. *lehmanniana*;
- development of prop roots at lower nodes on erect culms, for example in *Hyparrhenia tamba*;
- a protective casing around the roots, for example in *Stipagrostis ciliata* var. *capensis*.

Culms

The culms or stems of a grass plant are characterized by internodes interrupted by thickened nodes. The culms have the important function of bearing the leaves and inflorescences, and the branching pattern and posture (erect or oblique, for example) of the culms give a species its characteristic appearance. The position of the leaves with respect to the nodes also affects the effectiveness of photosynthesis. The position of the inflorescence, which is determined

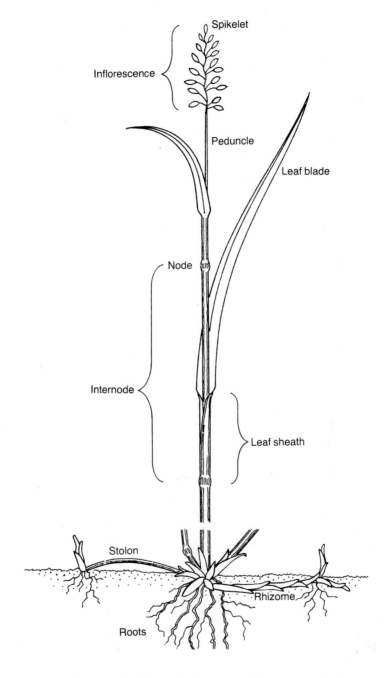

Figure 18. General structure of the grass plant.

by the length and growth form of the culm, is important for effective seed dispersal. The culms are also responsible for the transport of water and nutrients to and from the various parts of the grass plant. The following characteristics of the culm can be used to identify a species:
- length of the culm;
- position of the leaves and inflorescences on the culm;
- extent of branching from the nodes;
- growth form of the culm, for example erect, oblique or geniculate;
- nodes of the culm. In some species the nodes are dark, for example in *Stipagrostis obtusa*, or have a ring of hair, as in *Sorghum versicolor*.

Stolons
Stolons are modified culms growing horizontally on the soil surface and are important vegetative reproductive structures as roots, leaves, culms or inflorescences arise from their nodes. The presence or absence of stolons is an important distinctive feature in the identification of a grass species.

Rhizomes
Rhizomes are modified culms growing horizontally below the soil surface. The leaves of a rhizome are modified and reduced to membranous cataphylls (scale-like bracts). Rhizomes, like stolons, develop roots, culms, leaves or inflorescences at the nodes, thereby giving rise to new independent plants. Apart from vegetative reproduction, rhizomes also enable the plant to survive harsh winters and drought by means of these underground structures. The presence or absence of a rhizome as well as its length is important in the identification of a grass species.

Leaves
Leaves are lateral appendages of the culms and originate at the nodes (Figure 19). Chlorophyll gives the leaves their green colour which ranges from dark-green, light-green to blue-green. The most important function of the leaves is to produce organic compounds through the process of photosynthesis, with the aid of chlorophyll and with sunlight as source of energy. Grass leaves can be divided into the leaf blade or lamina, leaf sheath and ligule.

Leaf blade. In grasses, the leaf blade (lamina) is parallel-veined and usually long and narrow. Besides these similarities there are differences which can be used in identification (Figure 20). The following differences are important:
- different leaf blade shapes, for example linear, lanceolate or oval;
- leaf blade surface, ranging from flattened, folded, rolled or thread-shaped;
- differences in hairiness, ranging from densely pubescent to glabrous;

- the leaf margin may be rough, smooth, undulate, ciliate or thickened;
- the leaf tip may be acute, obtuse, rounded or rolled;
- the leaf blade of some species may be characteristically folded, as in *Setaria megaphylla*;
- some characters of the leaf blade are often affected by the age of the plant and by climatic factors such as rainfall and temperature.

Leaf sheath. The leaf sheath is really the leaf base which has developed into a sheath wrapped around the culm. The main function of the leaf sheath is to protect the growth zone of the internode and to support the internode as a whole.

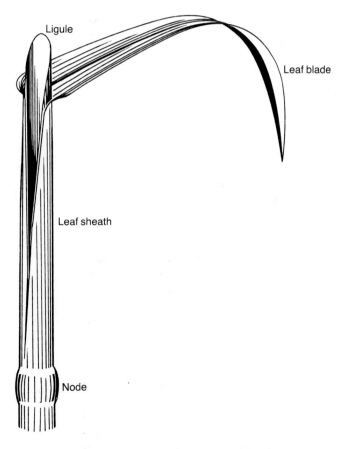

Figure 19. Tipical structure of the leaf blade.

The following differences in the leaf sheath may be used in the identification of a grass species:
- differences in the hairiness of the leaf sheath and of its margins which may range from densely pubescent to glabrous;
- differences in shape, ranging from rounded to strongly compressed;
- in some species the basal leaf sheaths are fan-shaped, overlapping and compressed, as in *Eustachys paspaloides*;
- basal leaf sheaths may become split into fibres, as in *Loudetia simplex*.

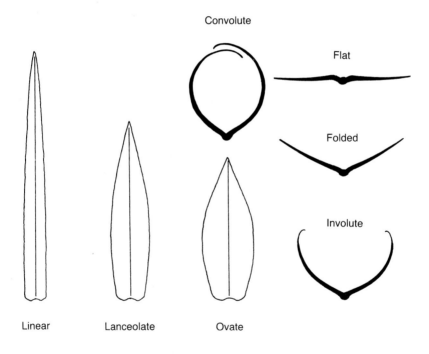

Figure 20. Different forms of the leaf blade.

Ligule. The ligule of the leaf occurs against the culm, at the junction of sheath and blade. The ligule is generally uniform in a genus. The following differences with regard to the ligule may be used in the identification of a grass (Figure 21):
- shape of the ligule, for example a ring of hairs, a membrane or a membrane fringed with hairs; in exceptional cases, the ligule may be a split membrane, as in *Trachypogon spicatus*, or a curved membrane, as in *Schizachyrium sanguineum*.
- length of the ligule;
- absence of a ligule, for example in *Echinochloa colona*.

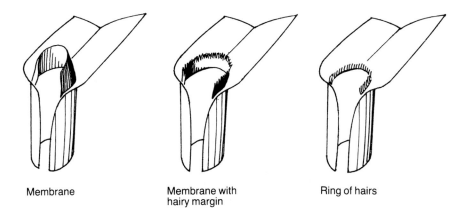

Membrane   Membrane with   Ring of hairs
                  hairy margin

Figure 21. Differences in the ligule.

REPRODUCTIVE PARTS

Inflorescences

The inflorescences of a grass plant is a flower-bearing structure borne at the tip of the culm. The difference between an inflorescence and a flowering branch is that after fruiting, no further vegetative growth can take place in an inflorescence. Morphologically, inflorescences can be divided into three groups, namely spikes, racemes and panicles (Figure 22).

Spikes are unbranched inflorescences, with the central axis bearing sessile spikelets.

Racemes are unbranched inflorescences, with the central axis bearing pedicellate (stalked) spikelets.

Panicles are branched inflorescences with a main axis giving rise to secondary axes which bear the spikelets.

The various inflorescence types can be distinguished by taking into account the morphological structure and the appearance of an inflorescence. For identification purposes, the inflorescence type is used more commonly than any other part of the grass plant. It therefore features prominently in identification systems.

The following features of the inflorescence may be used in the identification of a grass species:

- inflorescence group and type;
- length and width of the inflorescence;
- arrangement of the spikelets;
- arrangement of the inflorescence axes;
- inflorescences of some grasses, for example *Aristida bipartita*, break off at maturity, to be dispersed by the wind.

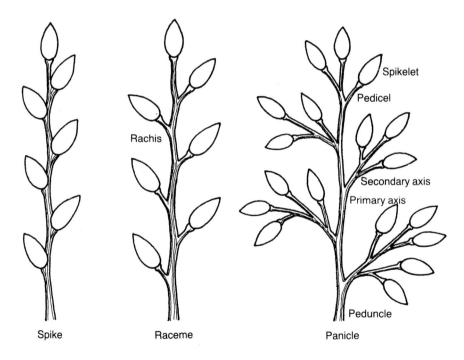

Figure 22. The three inflorescence types.

Spikelets

A spikelet is the basic flower-bearing unit of an inflorescence and consists of a rachilla, two glumes and one or more florets. Each floret consists of the lodicules, stamens and an ovary enveloped by the lemma and palea (Figure 23). The main function of the spikelet is the protection and dispersal of grain (seed). Spikelets are usually also responsible for the effective germination of the grain. Although there are remarkable structural similarities between spikelets, there are also external differences that are often used to distinguish between different genera and species, for example:

- length of the spikelet;
- number of florets per spikelet;
- general shape of the spikelet;
- presence and shape of awns and bristles;
- presence and length of the pedicels (stalks);
- position and shape of spikelet parts such as the glume, lemma and palea;
- hairiness and position of hairs on the various spikelet parts.

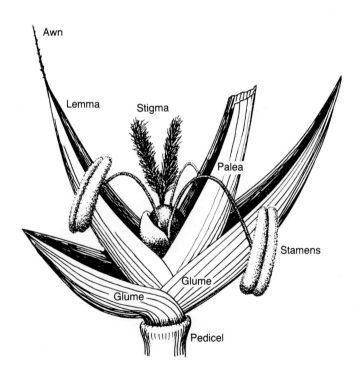

Figure 23. Tipical parts of an one-floret spikelet.

IDENTIFICATION SYSTEM

The inflorescence of a grass is often used as an aid in the identification of grasses, because it seldom varies within a species. For the purpose of this guide, the inflorescences of the species concerned have been divided into four groups based on their general appearance and morphological characteristics: solitary, digitate, paniculate and inconspicuous inflorescences (also see colour codes). These groups can be divided further into twelve inflorescence classes as shown in Figure 24.

When a grass has to be identified, the inflorescence should be studied carefully. By comparing the inflorescence with the sketches in figure 24, it can be placed in one of the inflorescence classes. With the aid of the page references and the photographs of the inflorescences, the inflorescence of the unknown plant can be compared with those in the same class. One or more species of which the inflorescences match that of the plant concerned, can then be tentatively chosen. Finally, the description and photographs are used to identify the plant in question. If this is impossible, another inflorescence class should be chosen and the identification process repeated. It may also be that the plant belongs to a relatively uncommon species not treated in this book. For the study and comparison of the spikelets, a hand lens of 10x magnification should be used.

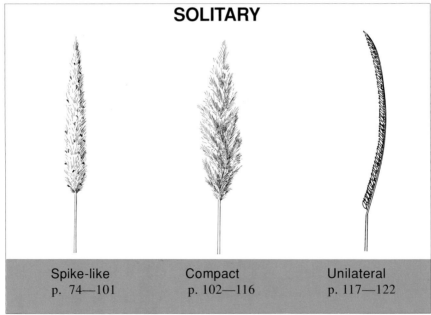

Figure 24. Subdivision of the four inflorescence groups.

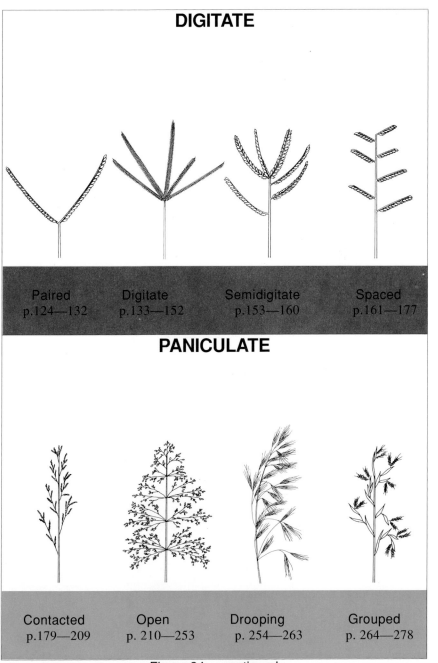

Figure 24. continued

INCONSPICUOUS

p. 280

LITERATURE CONSULTED

GIBBS RUSSELL, G.E., WATSON, L., KOEKEMOER, N., SMOOK, L., BARKER, N.P., ANDERSON, H.M. & DALLWITZ, M.J., 1990. Grasses of southern Africa. **Memoirs of the Botanical Survey of South Africa** no. 58.
HäFLIGER, E. & SCHOLZ, H. 1980. **Grass weeds,** Vol. 1 Ciba Geigy, Basel, Switzerland.
REYNEKE, W.F., COETZER, L.A. & GROBBELAAR, N. 1987. **Plantkunde - morfologie en sitologie,** 2nd edn. Butterworth, Durban.
VAN DER SCHIJFF, H.P. 1987. **Algemene plantkunde,** 5th edn. J.L. van Schaik, Pretoria.

SOLITARY INFLORESCENCES

SPIKE-LIKE	COMPACT	UNILATERAL
p. 74—101	p. 102—116	p. 117—122

ANTHEPHORA ARGENTEA

Silver Wool Grass
Silwerborseltjiegras

A tufted perennial with a short rhizome and culms up to 1,0 m tall. **Inflorescence** a silver woolly spike up to 150 mm long and 5 mm wide. Flowers from November to April. **Spikelets** up to 6 mm long, hairy, with an acute lower glume. **Leaf blade** up to 3 mm wide, stiff and usually folded. **Leaf sheath** glabrous and round. **Ligule** a conspicuous membrane, up to 8 mm long, sometimes split.

Habitat The grass is normally limited to Kalahari thornveld. Grows on loose sandy soil, often on sand dunes. Prefers undisturbed veld. **Biomes**: Savanna and Nama-Karoo.

General A valuable and palatable pasture grass with a high nutritive value. When abundant in natural veld, it can be regarded as an indicator of good veld management. *Anthephora argentea* is often confused with *Anthephora pubescens*, but the latter is distinguished by its thicker inflorescence (up to 10 mm wide), its thickened leaf margin and curled old leaves. Although it rarely shares the same habitat, *Anthephora argentea* is sometimes confused with *Elionurus muticus*, but the spikelets of the latter occur in pairs, with the inflorescence typically sickle-shaped when mature. **Grazing value** very high. **Ecological status**: Decreaser.

ANTHEPHORA PUBESCENS

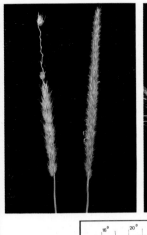

Wool Grass
Borseltjiegras

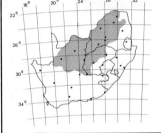

A tufted perennial with a short rhizome and unbranched culms up to 1,5 m tall. **Inflorescence** a hairy spike, up to 150 mm long and 10 mm wide, varying from white and straw-coloured to grey. Flowers from December to April. **Spikelets** up to 8 mm long, densely hairy and lower glumes generally tapered or with a short awn. **Leaf blade** up to 5 mm wide, glabrous or hairy, flat, leaf margin sinuate, old leaves curled. **Leaf sheath** usually round, glabrous or hairy. **Ligule** a papery membrane, up to 6 mm long.

Habitat Widespread, but more common in relatively dry areas (with annual rainfall below 650 mm). Usually occurs in undisturbed veld. Grows on most soil types, but prefers sandy soils. **Biomes**: Grassland, Savanna and Nama-Karoo.

General A very palatable pasture grass well utilized by grazers. Although it has a relatively low leaf production, a large percentage of the yield is utilized because it is so highly palatable. Planted as pasture with success in arid regions. The grass is drought-resistant and performs well on soils with a low nutritional status. Where abundant in natural veld, it indicates good veld conditions. **Grazing value** very high. **Ecological status**: Decreaser.

CENCHRUS CILIARIS

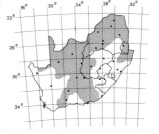

Blue Buffalo Grass
Bloubuffelgras

A tufted, shrubby perennial with culms up to 1,0 m tall, often geniculate and branched, usually with roots at the lower nodes. **Inflorescence** a dense, hairy, cylindrical spike, up to 100 mm long. Flowers from August to April. **Spikelets** up to 5 mm long and surrounded basally by numerous wavy bristles up to 10 mm long. **Leaf blade** up to 8 mm wide, bright green or blue-green, usually glabrous, or with scattered hairs, particularly near the ligule. **Leaf sheath** keeled. **Ligule** a ring of short hairs.

Habitat Common in warm and dry parts, but also widespread in other regions, where it is dispersed by man. Grows on all types of soil, with a preference for sandy, limy and stony soils. **Biomes**: Savanna, Grassland and Nama-Karoo.

General A palatable species with a high leaf production. Various cultivars are available and are generally planted as pasture in warm and dry areas. The grass is suitable for hay. However, it becomes hard and fibrous in winter and late in the growing season. *Cenchrus ciliaris* may be confused with *Enneapogon cenchroides* and *Enneapogon scoparius*, which often occur in the same habitat, and with certain species of *Setaria*. *Enneapogon cenchroides* and *Enneapogon scoparius* have spike-like, contracted panicles and no bristles at the base of the spikelets. In species of *Setaria* the spikelets drop from the spike without the bristles, whereas in *Cenchrus ciliaris* the bristles are attached to the spikelets. **Grazing value** mostly high. **Ecological status**: Decreaser.

ELIONURUS MUTICUS

Wire Grass
Koperdraad

A dense, tufted, perennial grass with culms up to 1,0 m tall. **Inflorescence** a single, upright, silver, woolly spike up to 120 mm long. Mature inflorescence is white and curled in a typical sickle-shape. Flowers from September to May. **Spikelets** occurring in pairs, of which one member is pedicellate and without an awn, and the other sessile, up to 14 mm long and the lower glume is often divided into 2 teeth up to 7 mm long.
Leaf blade narrow, thread-like and inrolled, hairy or glabrous, colour changes from dark green to copper-brown late in the season. Leaves have a bitter smell and taste. **Leaf sheath** thin, round and glabrous. **Ligule** a dense ring of hairs.

Habitat Commonly occurs in open grassland in higher rainfall areas and in broken country of lower rainfall areas. Grows on many kinds of soils, but prefers poor, stony soil. **Biomes**: Savanna and Grassland.

General A very unpalatable grass which is utilized mainly in spring because it is one of the first grasses to sprout. It can be viewed as an indicator of poor veld management when abundant in veld. The inflorescence may be confused with those of *Anthephora argentea*, *Anthephora pubescens*, *Digitaria monodactyla* and *Schizachyrium jeffreysii*, but can be distinguished by its typical thread-like leaf blades and sickle-shaped mature inflorescences. **Grazing value** very low. **Ecological status**: mostly Increaser IIb.

FINGERHUTHIA AFRICANA

Thimble Grass
Vingerhoedgras

A tufted perennial (rarely an annual) with unbranched culms up to 0,9 m tall. **Inflorescence** a dense cylindrical spike, up to 50 mm long. Flowers from September to May. **Spikelets** up to 5,5 mm long, flattened, with glumes bearing short awns and short hairs on the keels. **Leaf blade** up to 4 mm wide, flattened, sometimes folded, with a few scattered hairs. **Ligule** a ring of short hairs.

Habitat   Widespread on well drained sandy and gravelly soils. Often found on limestone outcrops and disturbed sites. Prefers warm and sunny localities. **Biomes**: Savanna, Grassland, Nama-Karoo and Succulent Karoo.

General   A relatively palatable grass with a low leaf production; becomes hard with age. May be confused with particularly *Fingerhuthia sesleriiformis* and *Pennisetum sphacelatum*. They are however, dense, tufted species with longer inflorescences and are limited to clayey soils in vleis and near rivers. **Grazing value** average. **Ecological status**: variable but mostly Decreaser.

HETEROPOGON CONTORTUS

Spear Grass
Assegaaigras

A tufted perennial with culms up to 0,7 m tall. **Inflorescence** a terminal spike-like raceme up to 70 mm long, sparsely or densely hairy and bearing conspicuous long brown awns which tend to cluster when ripe. Flowers from October to March. **Spikelets** of two types, one awnless and the other with a hairy brown awn. **Leaf blade** up to 7 mm wide, variable with regard to hairiness and shape, and the lower leaves usually with blunt tips. **Leaf sheath** compressed and keeled. **Ligule** an inconspicuous membrane.

Habitat Widespread in open grassland and bushveld areas. Common on well drained, stony soils. Generally found on stony slopes and disturbed soils such as roadsides, where it can form thick stands. **Biomes**: Grassland, Savanna, Nama-Karoo and Fynbos.

General A relatively good, hardy and fast-growing pasture grass, with a declining grazing value as the season progresses. The long awns can create problems by penetrating the skin of livestock, thus decreasing the quality of the skin, wool and meat. *Heteropogon contortus* can be confused with *Trachypogon spicatus*, but is distinguished by the shorter culms of the former and the ring of short white hairs around the nodes of *Trachypogon spicatus*. Preferred by mountain zebra, roan antelope and waterbuck. **Grazing value** average to high. **Ecological status**: Variable.

*HORDEUM MURINUM subsp. MURINUM

False Barley
Wildegars

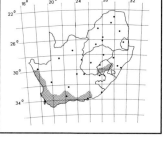

A tufted annual with culms up to 0,50 mm tall, seldom upright. **Inflorescence** a dense, prickly, flattened spike, up to 110 mm long. Flowers from September to October. **Spikelets** in groups of three which are arranged on either side of the primary axis, each spikelet with conspicuous awns, up to 40 mm long. **Leaf blade** up to 8 mm wide, flattened, soft and hairy or glabrous. **Leaf sheath** membranous on the sides, rounded and usually hairy, upper part usually partially encloses the inflorescence. **Ligule** a white membrane.

Habitat Occurs mainly on disturbed sandy soils such as roadsides, gardens and fields. **Biomes**: Savanna, Fynbos and Succulent Karoo.

General Indigenous to South-east Asia and to southern and western Europe, but now widespread, also occurring in South Africa. The grass is often found as a weed in lucerne, lowering the purity of the hay. Apparently causing ulcers in especially horses' mouths.

IMPERATA CYLINDRICA

Cottonwool Grass
Donsgras

A perennial grass spreading by means of strong rhizomes, with unbranched culms up to 1,2 m tall. **Inflorescence** a dense, cylindrical, spike-like panicle up to 250 mm long. Flowers from August to June. **Spikelets** up to 6 mm long, in pairs, pedicellate and covered with long, silky, white hairs. **Leaf blade** up to 12 mm wide, hard, with a distinct midrib and a hard, sharp tip. **Leaf sheath** round and smooth. **Ligule** an inconspicuous membrane with scattered hairs.

Habitat Common in poorly drained, damp soils such as vlei areas and riverbanks, where it can form dense stands. Also in open grassland with a high rainfall. **Biomes:** Grassland, Savanna and Fynbos.

General Poorly utilized by grazers owing to the hardness of its leaves. It is resistant to seasonal veld fires. Difficult to control in fields because of its vigorously growing rhizomes. Important in erosion control, particularly on riverbanks. The rhizome is apparently eaten by herdsmen in Lesotho. Used as thatching grass in Mozambique. Preferred by reedbuck. **Grazing value** very low to low. **Ecological status**: mostly Increaser I.

* LOLIUM PERENNE

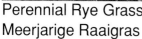

Perennial Rye Grass
Meerjarige Raaigras

A tufted perennial up to 0,9 m tall. **Inflorescence** a flattened spike, up to 300 mm long. Flowers from October to April. **Spikelets** up to 15 mm long, flattened, oblong and glabrous, each bearing up to 12 'seeds', and usually without an awn. **Leaf blade** up to 4 mm wide, glabrous, glossy and flat, young leaves usually rolled. **Leaf sheath** a round structure. **Ligule** an inconspicuous glabrous membrane.

Habitat Grows on a wide variety of soils, usually occurring along roadsides, in uncultivated lands and other disturbed areas. Prefers high-rainfall regions with a cool climate. **Biomes**: Grassland, Fynbos, Savanna and Nama-Karoo.

General Originally brought from Europe. It is planted as a perennial winter pasture. The grass responds well to irrigation, and under ideal conditions it can provide palatable grazing with a high crude protein content. Can be poisonous when infected by a poison producing bacterium. Can also be a troublesome weed in areas with ideal climatic conditions. *Lolium perenne* cross freely with other *Lolium* species. Can be distinguished from *Lolium multiflorum* by the presence of awns (up to 15 mm long) on the 'seeds' (lemmas) of the latter. **Grazing value** very high under ideal growing conditions.

MERXMUELLERA DISTICHA

Mountain Wire Grass
Bergkoperdraad

A dense tufted perennial with erect, unbranched culms up to 0,7 m tall. **Inflorescence** a flattened spike, up to 100 mm long, consisting of two rows of spikelets. Flowers from October to May. **Spikelets** up to 18 mm long (awns included), flattened, glabrous and glumes almost equal. **Leaf blade** up to 3 mm wide, filiform and usually glabrous. **Leaf sheath** usually glabrous. **Ligule** a ring of short white hairs.

Habitat Common in high-lying mountainous short grassland, but adapted to a wide range of habitats, from coastal regions to vleis and marshes in mountainous grassland. **Biomes**: Nama-Karoo, Grassland and Fynbos.

General An unpalatable grass with tough leaves, which is poorly utilized by livestock. Its abundance in natural veld in certain Karoo mountain veld areas can be seen as an indicator of poor veld management. **Grazing value** very low to low. **Ecological status**: Increaser IIb.

*PENNISETUM SETACEUM

Fountain Grass
Pronkgras

A tufted perennial, with culms up to 1,0 m tall. **Inflorescence** a hairy purple spike up to 250 mm long. Flowers from November to July. **Spikelets** up to 2,5 mm long, surrounding bristles up to 20 mm long, short pedicels covered with dense white hairs. **Leaf blade** up to 2 mm wide, linear, with a very scabrid leaf margin. **Leaf sheath** round to slightly keeled. **Ligule** a membrane with a hairy margin.

Habitat Usually occurs in disturbed dry areas, often on steep stony slopes, along roads and at excavations. **Biomes**: Savanna, Fynbos and Nama-Karoo.

General A popular ornamental grass, particularly in grass gardens. Introduced from northern Africa. Of little grazing value, owing to its hard scabrid leaves. Related to *Pennisetum villosum*, another ornamental, which is a smaller plant with a shorter and broader, light green to white inflorescence and smooth leaves.

PENNISETUM SPHACELATUM

False Bristle Grass
Bulgras

A tufted perennial with a short rhizome and culms up to 1,0 m tall. **Inflorescence** a dense, soft spike up to 150 mm long, often with hairs on the culm directly below the inflorescence. Flowers from November to April. **Spikelets** approximately 3 mm long, the surrounding bristles as long, or twice as long as the spikelet. **Leaf blade** up to 3 mm wide, flattened or filiform. **Ligule** a ring of hairs.

Habitat Generally in wet places such as vleis and other drainage areas, usually on deep heavy soils, but also on other moist soils. Often found in mountainous areas, and sometimes in shade. **Biome**: Grassland.

General A relatively palatable grass with a high leaf production. However, later in the growing season it becomes hard and fibrous and less acceptable as grazing. Plays an important role in the stabilization of watercourses. Sometimes confused with the closely related *Pennisetum macrourum*, but the latter is a much larger plant (up to 2,5 m tall) and its inflorescences are longer (120—250 mm) than those of *Pennisetum sphacelatum*. Also confused with species of *Setaria*, but in *Setaria* the bristles remain on the inflorescence when the 'seeds' disperse, but not in species of *Pennisetum*. **Grazing value** average. **Ecological status**: Decreaser.

* PENNISETUM VILLOSUM

Feather Top
Haarwurmgras

A tufted perennial with a rhizome, culms up to 0,6 m tall. **Inflorescence** a dense feathery spike, up to 100 mm long. Flowers from January to May. **Spikelets** up to 14 mm long, with surrounding feathery bristles up to 70 mm long. **Leaf blade** up to 3 mm wide, linear, flattened or rolled, glabrous or with a few scattered hairs. **Leaf sheath** compressed or keeled, usually with hairs on the margins. **Ligule** a ring of hairs.

Habitat   Occurs in disturbed areas, often along roads and pavements. **Biomes**: Grassland and Savanna.

General   Originally from Ethiopia. Used as an ornamental grass and is particularly popular in grass gardens. Closely related to *Pennisetum setaceum*, another ornamental grass with a purple cylindrical spike up to 250 mm long.

PEROTIS PATENS

Cat's Tail
Katstertgras

A tufted annual to weak perennial with culms up to 0,6 m (rarely taller) **Inflorescence** a soft, straight spike of approximately uniform width and up to 300 mm long. Flowers from November to April. **Spikelets** up to 3 mm long, bears long, delicate awns, up to 22 mm long. **Leaf blade** up to 12 mm wide, relatively short with a sharp tip, flattened and wavy, glabrous, except for the distinctly spaced hairs along the leaf margin. **Leaf sheath** glabrous and rounded. **Ligule** an inconspicuous membrane.

Habitat Occurs in disturbed areas, usually on poor, sandy soils. Often in dry, bare patches. **Biomes**: Grassland and Savanna.

General Can be regarded as an indicator of poor veld management when abundant in natural veld. Often a weed in disturbed areas where the soil is poor. An attractive grass which is sometimes used in flower arrangements. **Grazing value** very low. **Ecological status**: Increaser IIc.

SCHIZACHYRIUM JEFFREYSII

Silky Autumn Grass
Harige-herfsgras

A tufted perennial often with a short rhizome and branched culms up to 1,0 m tall. **Inflorescence** consists of numerous single racemes arranged in a false panicle, up to 80 mm long, each conspicuously cover-ed with fine white hairs. Flowers from February to June. **Spikelets** up to 8 mm long, paired, one member sessile with a bent awn up to 16 mm long, and the other pedicellate with a short straight awn. **Leaf blade** up to 5 mm wide, generally glabrous, flat, reddish. **Leaf sheath** round. **Ligule** a short membrane.

Habitat Generally in stable veld in bushveld areas with a moderate rainfall. In sandveld areas it prefers wet localities such as vleis and riverbanks. Sometimes found in watercourses along roads. **Biome**: Savanna.

General An unpalatable species with a low grazing value, but with a relatively high production. Fast losing its palatability as it matures. Sometimes used as thatching grass. The inflorescence may be confused with that of *Elionurus muticus*, but the latter generally has fewer inflorescences per culm and is a dense, tufted grass with narrow thread-shaped leaves (up to 2 mm wide). Preferred by sable antelope.

SCHIZACHYRIUM SANGUINEUM

Red Autumn Grass
Rooiherfsgras

A tufted perennial sometimes with short rhizome, culms up to 1,2 m tall, with repeated branching from the upper nodes. **Inflorescence** a false panicle consisting of a number of single racemes up to 100 mm long. Flowers from January to May. **Spikelets** of two types and in pairs, of which one member is pedicellate, with a short, straight awn, and the other is sessile, up to 9 mm long with a longer, bent awn. **Leaf blade** up to 7 mm wide and glabrous, with a distinct midrib. **Leaf sheath** smooth and glabrous. **Ligule** a strongly curved membrane.

Habitat Widespread in open grassland and in open patches in bushveld. Limited to edges of vleis, riverbanks and moist soils in lower rainfall regions. Grows on almost any type of soil. **Biomes:** Grassland and Savanna.

General A reasonably good pasture grass when young, but soon loses its palatability when it becomes hard and fibrous at the flowering stage, then poorly utilized by grazers. When few or no other thatching grasses are available, this grass is used for thatching. In the vegetative stage (when inflorescences are absent) *Schizachyrium sanguineum* could be confused with *Heteropogon contortus* or certain strains of *Themeda triandra*, with which it often shares its habitat. However, in *Heteropogon contortus* the ligule is a slightly curved membrane and in *Themeda triandra* a split membrane. Preferred by roan antelope. **Grazing value** low to average. **Ecological status**: Increaser I.

89

SEHIMA GALPINII

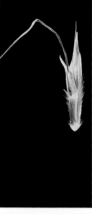
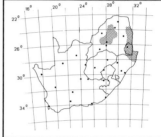

Yellow Turf Grass
Geelturfgras

A tufted perennial with unbranched culms up to 1,8 m tall. **Inflorescence** a solitary spikelike raceme up to 190 mm long, borne terminally on the culm. Flowers from October to April. **Spikelets** occur in pairs of which one member is sessile and up to 15 mm long, with a bent and contorted awn up to 40 mm long, while the other is pedicellate and awnless. **Leaf blade** up to 6 mm wide, glabrous, flattened and scabrid on the surface, with a thickened leaf margin. **Leaf sheath:** basal sheaths hairy. **Ligule** a ring of hairs.

Habitat Generally in open patches in bushveld. Often grows on black peat soils, where it may form dominant stands. **Biome:** Savanna.

General Grazing value unknown, but it is probably a palatable and relatively valuable grass. It is sometimes used as thatching grass. *Sehima galpinii* is sometimes confused with *Ischaemum afrum*, *Urelytrum agropyroides* and *Trachypogon spicatus*. However, the ligules of those three species are membranous, whereas that of *Sehima galpinii* is a ring of hairs.

SETARIA INCRASSATA

Vlei Bristle Grass
Vleimannagras

A tufted perennial, with a creeping rhizome and culms up to 2,0 m tall, usually with hairy nodes. **Inflorescence** spikelike, up to 100 mm long, often tapering. Flowers from October to May. **Spikelets** up to 3 mm long, without a black tip, lower glumes three-veined. **Leaf blade** up to 14 mm wide, flat and usually blue-green.

Habitat Generally in wet areas such as vleis and marshes. Found mostly on black clay soils. Also grows on riverbanks, stony slopes and forest edges. **Biomes:** Savanna, Grassland, Nama-Karoo and Fynbos.

General A palatable species with a moderate leaf production. Confused with the closely related *Setaria nigrirostris* which often shares the same habitat. *Setaria nigrirostris* can be distinguished by its more strongly developed, branched rhizome, and the dark purple or black tips of its spikelets. Preferred by gemsbok. **Grazing value** average to high. **Ecological status**: Decreaser.

SETARIA NIGRIROSTRIS

Black-seed Bristle Grass
Swartsaadmannagras

A tufted perennial with a well-developed brown rhizome and unbranched culms up to 1,2 m tall. **Inflorescence** spikelike, up to 100 mm long, consisting of densely packed and relatively large spikelets with bristles in between. Flowers from October to April. **Spikelets** up to 5 mm long with characteristic dark tips. **Leaf blade** up to 10 mm wide, flat, dark green and glossy. **Leaf sheath** hairy. **Ligule** a ring of hairs.

Habitat Generally occurs in open grassland or open patches in bushveld, on black peat soils. Often found on riverbanks or in other moist, low lying areas. **Biomes**: Savanna and Grassland.

General A palatable grass with a reasonably good leaf production, well utilized, but less acceptable for grazing when mature. *Setaria nigrirostris* is easily confused with the closely related *Setaria incrassata*, but the inflorescence of the latter tapers to a sharp tip, the spikelets are smaller (up to 3 mm long) and do not have the characteristic dark tips, and the leaves are not dark green. Preferred by blesbok. **Grazing value** high. **Ecological status**: Decreaser.

SETARIA PALLIDE-FUSCA

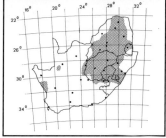

Garden Bristle grass
Tuinmannagras

A tufted annual, sometimes with branched culms up to 0,9 m tall. **Inflorescence** spike-like, up to 100 mm long. **Spikelets** up to 3 mm long, glabrous and surrounded by yellow, orange to dark purple bristles. **Leaf blade** up to 8 mm wide, relatively short, flattened and glabrous, except for hairs close to the ligule. **Leaf sheath** compressed and glabrous. **Ligule** membranous, with hairs.

Habitat Occurs on most soil types. Common in disturbed areas, particularly in gardens and in cultivated and fallow lands. Prefers positions where water accumulates. **Biomes**: Savanna, Grassland, Nama-Karoo and Succulent Karoo.

General A palatable grass which is fairly well grazed and provides high quality hay, but unfortunately it has a low leaf production. A common weed in cultivated fields, which can be controlled by tilling the soil at the seedling stage. Closely related to *Setaria ustilata* which has bi-coloured bristles, coarsely rugose lemmas and usually occurs in shade. **Grazing value** low. **Ecological status**: Increaser IIc.

SETARIA SPHACELATA var. SERICEA

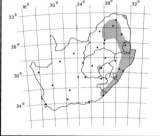

Golden Bristle Grass
Goue-mannagras

A tufted perennial, occasionally with a creeping rhizome, with culms up to 2,0 m tall. **Inflorescence** dense and spike-like, 100—300 mm long. Flowers from October to May. **Spikelets** glabrous and up to 4 mm long. **Leaf blade** up to 15 mm wide, flattened or folded, hairy or glabrous. **Leaf sheath** round, or lower ones flattened. **Ligule** a ring of hairs.

Habitat Common on wet soils, with a preference for clayey wet soils. Often in vleis and spongy areas. Generally in uncultivated lands and other disturbed areas with wet soil. **Biomes**: Grassland and Savanna.

General A good pasture grass with a high palatability and leaf production. Selected varieties such as 'Kazungula' are planted in high rainfall regions, particularly on wet clay soils. However, it is not drought-resistant and is unable to withstand continuous grazing. Particularly suitable for hay-making. *Setaria pallide-fusca* may be confused with *Setaria sphacelata* var. *sericea*, but the former can easily be distinguished by its annual growth form and culms which are shorter than 0,9 m. **Grazing value** high. **Ecological status**: Decreaser.

SETARIA SPHACELATA var. SPHACELATA

Common Bristle Grass
Gewone Mannagras

A tufted perennial with a short rhizome and culms up to 1,0 m tall. **Inflorescence** spike-like, up to 150 mm long, with yellow to reddish brown bristles between the spikelets. Flowers from October to June. **Spikelets** up to 3,5 mm long, with wrinkled lemmas. **Leaf blade** up to 5 mm wide, glabrous to densely hairy, rolled or flattened. **Leaf sheath** glabrous to densely hairy. **Ligule** a ring of short hairs.

Habitat Adapted to a wide range of habitats, which may vary from riverbanks and other damp areas to stony slopes. Generally grows on well-drained sandy soil. **Biomes**: Savanna, Grassland and Fynbos.

General A palatable species which is well utilized. Is confused with the closely related *Setaria sphacelata* var. *sericea* which is a larger plant (up to 2 m tall), with wider leaves (up to 10 mm). **Grazing value** high. **Ecological status**: mostly Decreaser.

SETARIA SPHACELATA var. TORTA

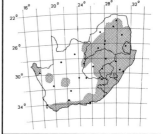

Creeping Bristle Grass
Kruipmannagras

A tufted perennial with a creeping rhizome, unbranched culms up to 0,5 m tall, and occasionally rooting from the lower nodes. **Inflorescence** spike-like, up to 80 mm long, with purple or sometimes yellowish bristles between the spikelets. Flowers from September to March. **Spikelets** up to 3 mm long, solitary or in groups of 2 or 3, each group with 7—15 lax, scabrid bristles. **Leaf blade** up to 3 mm wide, keeled, generally folded or involute, older leaves often contorted and curled. **Leaf sheath:** basal sheaths keeled and fan-shaped. **Ligule** a ring of hairs.

Habitat Occurs in open bushveld and grassland, on plains and slopes, often in disturbed areas such as overgrazed veld and roadsides. Usually grows on well-drained soils, but sometimes also on moist soils. **Biomes**: Fynbos, Grassland, Savanna, Nama-Karoo and Succulent Karoo.

General A palatable grass, relatively well utilized. Plays an important role in soil conservation. *Setaria sphacelata* var. *torta* is the smallest representative of the *Setaria sphacelata* complex. **Grazing value** average. **Ecological status**: Variable.

SETARIA USTILATA

Shade Bristle Grass
Skadumannagras

A tufted annual, with erect or geniculate culms, up to 0,65 m tall. **Inflorescence** spikelike, often oval and sometimes up to five times as long as wide. Flowers from January to May. **Spikelets** up to 2,5 mm long, with 6—10 bi-coloured bristles per spikelet, lemmas coarsely rugose. **Leaf blade** up to 12 mm wide. **Leaf sheath** slightly flattened. **Ligule** a ring of short white hairs.

Habitat Generally in the drier bushveld areas, growing mostly in the shade of trees and bushes, often in moist soil. **Biome**: Savanna.

General Probably of little grazing value because of its low leaf production. Confused with *Setaria pallide-fusca*, which has finely rugose lemmas and a inflorescence up to 10 times as long as wide.

TRACHYPOGON SPICATUS

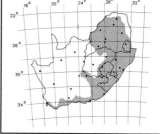

Giant Spear Grass
Bokbaardgras

A tufted perennial with culms up to 1,2 m tall and nodes with a ring of white hairs. **Inflorescence** a single raceme (rarely up to 5), up to 250 mm long, borne terminally on the culm. Flowers from January to May. **Spikelets** usually up to 8 mm long, hairy, with a velvety awn. **Leaf blade** up to 5 mm wide, generally rolled and hairy. **Leaf sheath** round. **Ligule** a tripartite membrane.

Habitat Common in open, undisturbed sour grassland, but also occurs in bushveld. Grows on most kinds of soil, with a preference for sandy soils. **Biomes**: Grassland, Savanna and Fynbos.

General Of medium palatability and fairly well utilized in the young stage. Becomes hard later with a consequent drop in grazing value. A good indicator of sour grassland. During certain stages *Trachypogon spicatus* may be confused with *Heteropogon contortus* or *Urelytrum agropyroides* but it can be distinguished by the ring of white hairs around its nodes. Preferred by reedbuck. **Grazing value** average to low. **Ecological status.** Increaser I.

TRAGUS BERTERONIANUS

Common Carrot-seed Grass
Gewone Wortelsaadgras

A tufted annual with short culms, up to 300 mm tall. Rooting common at lower nodes. **Inflorescence** a long dense spike up to 100 mm long. Flowers from November to May. **Spikelets** 2—3 mm long, covered with hooked spines, densely arranged around the rachis, drop easily when mature and tend to stick to clothes. **Leaf blade** up to 5 mm wide, flattened and lanceolate to oval, leaf margin fringed with sparse hairs. **Leaf sheath** rounded. **Ligule** an inconspicuous ring of hairs.

Habitat Common as a weed in disturbed places such as uncultivated lands and trampled, overgrazed areas. Grows on poor soil, and prefers sandy soils. Occurs on compacted soils where few other grass species grow. **Biomes**: Grassland, Savanna and Nama-Karoo.

General Hardly any grazing value because of its low leaf production. The hooked spines attached to the 'seeds' can present problems with wool-bearing sheep because they enter the fleece, thus lowering the quality of the wool. The occurrence of this grass in natural veld indicates veld retrogression. However, it plays an important role in control of soil erosion on erodible soils. Could be confused with *Tragus racemosus* which has longer and larger spikelets (up to 5 mm long). Young plants are occasionally utilized by springbok. **Grazing value** very low. **Ecological status**: Increaser IIc.

TRAGUS RACEMOSUS

Grootwortelsaadgras
Large Carrot-seed Grass

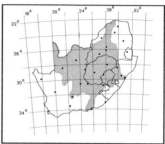

A tufted annual with culms up to 200 mm (rarely up to 400 mm) tall, which are usually geniculate, often rooting at the lower nodes. **Inflorescence** spike-like, up to 100 mm long, dense to slightly open. Flowers from November to May. **Spikelets** up to 5 mm long and in groups of 2—5, glumes with hooked hairs. **Leaf blade** up to 4 mm wide, with evenly spaced hairs on the margin. **Ligule** a membrane with a hairy margin.

Habitat Particularly prevalent in disturbed areas such as cultivated lands, overgrazed veld, and roadsides. Grows on sandy and limy soil. **Biomes**: Savanna, Grassland and Nama-Karoo.

General A grass that provides poor grazing owing to its low leaf production and short growth cycle. The bristly, prickly spikelets create problems with wool-bearing sheep because the spikelets become attached to the fleece, thus lowering its quality. Sometimes a weed in cultivated lands. *Tragus racemosus* is often confused with *Tragus berteronianus*, but can be distinguished by the inflorescence of the latter which is often denser and longer (up to 150 mm long) and the spikelets which are smaller (up to 4 mm long) than those of *Tragus racemosus*. **Grazing value** very low. **Ecological status**: Increaser IIc.

URELYTRUM AGROPYROIDES

Quinine Grass
Varkstertgras

A tufted perennial with unbranched culms up to 1,6 m tall. **Inflorescence** a single flattened raceme (sometimes two) up to 250 mm long, with conspicuous awns which typically bend outwards (like fish-bones) when mature. Flowers from October to June. **Spikelets** occur in pairs, of which one member is up to 8 mm long, sessile and awnless, while the other is pedicellate, with a distinctive flattened awn up to 120 mm long. **Leaf blade** up to 6 mm wide, with a distinct midrib, often rolled. **Leaf sheath** round. **Ligule** a conspicuous elongated membrane, often divided.

Habitat Occurs in grassland and open bushveld, often found on stony slopes. Generally grows in sandy soils. **Biomes**: Grassland and Savanna.

General A hard unpalatable grass which is poorly utilized by livestock. The leaves and culms have an bitter quinine-like taste (see English common name) which is characteristic of the species. The inflorescence of *Urelytrum agropyroides* is often confused with those of *Heteropogon contortus* and *Trachypogon spicatus*, although the latter two species both have hairy awns and the leaves and culms have no bitter taste. *Trachypogon spicatus* also has a ring of white hairs around the nodes and *Heteropogon contortus* is usually much shorter (up to 1 m tall). **Grazing value** very low. **Ecological status**: Increaser I.

ARISTIDA ADSCENSIONIS

Annual Three-awn
Eenjarige Steekgras

A tufted annual with culms up to 0,6 m tall. **Inflorescence** a contracted, spike-like panicle up to 250 mm long. **Spikelets** glabrous, with shades of purple, with a tripartite awn typical of most other species of *Aristida*. **Leaf blade** either glabrous or with sparse, scattered hairs. **Leaf sheath** glabrous and slightly flattened. **Ligule** an inconspicuous ring of hairs.

Habitat   Occurs mostly on disturbed and bare soils, where it is usually the first grass to start the succession process. **Biomes**: Savanna, Grassland and Nama-karoo.

General   The grass has little value for grazing because of its low leaf production and overall hardness. Where the grass occurs in natural veld, it is usually a indicator of overgrazing. Like most species of *Aristida*, it presents a problem for wool-bearing sheep and angora goats. The 'seeds' penetrate the fleece and thus lowers the quality of the wool and hair. **Grazing value** very low. **Ecological status:** Increaser IIc.

ARISTIDA CONGESTA subsp. CONGESTA

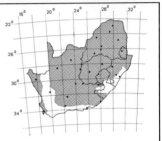

Tassel Three-awn
Katstertsteekgras

A tufted perennial (occasionally annual) with culms up to 0,9 m tall. **Inflorescence** a compact, spike-like panicle, up to 200 mm long, consisting of spikelets on short branches on the primary axis. Flowers from December to June. **Spikelets** each consisting of a single floret with a tripartite awn. The awn protrudes almost horizontally from the dense inflorescence. **Leaf blade** glabrous, initially flattened, but rolled when more mature. **Leaf sheath** keeled and glabrous or hairy. **Ligule** a ring of short hairs.

Habitat Common in bushveld and in open grassland on all soil types with a preference for loamy soil. Abundant along roadsides, on uncultivated lands and overgrazed veld. Occurs on hard, denuded soil where few other species can survive. **Biomes**: Savanna and Grassland.

General Of low grazing value as a result of its low leaf production and medium to low palatability. Grazed only in the young stage. Can, however, be palatable in semi-arid regions. Could be indicative of poor veld management. A useful pioneer grass for covering bare patches. Poses a problem for wool-bearing sheep and angora goats when 'seeds' penetrate the fleece and lower the quality of the wool and hair. Preference species for spingbok. **Grazing value** mostly very low. **Ecological status:** Increaser IIc.

103

ARISTIDA JUNCIFORMIS subsp. JUNCIFORMIS

Ngongoni Three-awn
Ngongoni-steekgras

A tufted perennial with culms up to 0,9 m tall. **Inflorescence** a narrow contracted panicle up to 200 mm long. Flowers from November to May. **Spikelets** up to 30 mm long (including awn), bearing a tripartite awn typical of most species of *Aristida*. **Leaf blade** up to 3 mm wide, generally inrolled and glabrous or slightly hairy. **Leaf sheath** usually round. **Ligule** inconspicuous.

Habitat Usually found in open mountain grassveld and in bushveld or fynbos with a relatively high rainfall. Prefers disturbed areas such as overgrazed veld and roadsides. Grows on most soil types, but prefers poor stony soils on slopes. **Biomes**: Grassland, Savanna and Fynbos.

General A very unpalatable grass, utilized by stock only to a limited degree and only at a young stage. The mature grass is tough and fibrous and unacceptable as pasture. However, the inflorescence can be of some use. Where the grass is dominant in the veld, it can be a good indicator of poor veld management in the past. The main problem with this grass is that where it invades natural veld, particularly high-rainfall mountainous grassveld, it is very difficult or virtually impossible to eliminate by normal grazing management practices. Form particularly dominant stands in parts of Natal (so-called Ngongoni veld) and Transkei. **Grazing value** very low. **Ecological status**: Increaser IIc.

ARISTIDA STIPITATA subsp. GRACILIFLORA

Long-awned Three-awn
Langnaaldsteekgras

A tufted perennial with culms up to 0,9 m tall. **Inflorescence** an open or contracted, sometimes drooping panicle, with relatively long, lax branches. Flowers from November to June. **Spikelets** up to 80 mm long (including awns), a long twisted column, which passes into three more or less equally long, prickly awns. **Leaf blade** up to 3 mm wide, glabrous and smooth, often rolled from the leaf margins. **Leaf sheath** round. **Ligule** a ciliate membrane.

Habitat Mostly in open bushveld on sandy to loamy soils. Often associated with heavily overgrazed veld, seepage lines and rocky outcrops or koppies. **Biome**: Savanna.

General A grass with little value for grazing owing to its general hardness and low leaf production. Can be regarded as an indicator of overgrazed veld. Can be distinguished from the closely related *Aristida stipitata* subsp. *stipitata* by the larger growth form and more compact, spike-like inflorescence of the latter. **Grazing value** very low. **Ecological status**: Increaser IIc.

ARISTIDA STIPITATA subsp. STIPITATA

Long-awned Three-awn
Langnaaldsteekgras

A tufted perennial with culms up to 1,5 m tall, often branching from the upper nodes. **Inflorescence** a dense, spike-like panicle, up to 300 mm long. Flowers from December to April. **Spikelets** up to 90 mm long (including awns), with a twisted column, passing into three rough, more or less equally long awns. **Leaf blade** up to 4 mm wide, glabrous, smooth and often rolled from the leaf margin. **Leaf sheath** round. **Ligule** a ciliate membrane.

Habitat   Common in open bushveld on deep sandy and sometimes limy soils. Often found in disturbed or stony and rocky areas, sometimes along vleis. **Biome**: Savanna.

General   With little value for grazing value owing to its general hardness and low leaf production. The grass can be regarded as an indicator of overgrazing. Can be distinguished from closely related *Aristida stipitata* subsp. *graciliflora* by the smaller growth form and more open inflorescence of the latter. **Grazing value** very low. **Ecological status**: Increaser IIc.

ARUNDINELLA NEPALENSIS

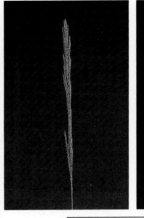

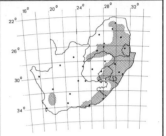

River Grass
Riviergras

A tufted perennial with a short, hard rhizome and unbranched, erect culms, up tot 1,5 m tall. **Inflorescence** a dense contracted panicle, up to 300 mm long, with a primary axis which is usually thicker than the secondary branches. Flowers from December to March. **Spikelets** up to 6 mm long, usually in pairs, glabrous, with an inconspicuous awn up to 6 mm long. **Leaf blade** up to 10 mm wide, hairy or glabrous, flat or rolled. **Ligule** a short membrane with a hairy margin.

Habitat   Usually grows in vleis, marshes, on riverbanks and often also in wet grassland. **Biomes**: Grassland and Savanna.

General   A palatable but tough grass, which is well utilized by stock. It is sometimes used as thatching grass. Of medicinal use in Lesotho.

CENTROPODIA GLAUCA

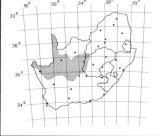

Gha Grass
Ghagras

An annual to weak perennial with short rhizomes, which are covered with papery bracts, culms up to 1,0 m tall and particularly brittle and fragile. **Inflorescence** a dense contracted panicle up to 120 mm long. Flowers from September to May. **Spikelets** up to 10 mm long, glumes up to 10 mm long, with bent awns up to 8 mm long. **Leaf blade** up to 10 mm wide, flattened and scabrid, with a fine sharp tip. **Leaf sheaths** round, basal sheaths generally densely hairy. **Ligule** a ring of short hairs.

Habitat Occurs mainly in sandveld in deep sandy soils or on sand dunes. More common on red sand and sometimes grows on gravelly soils. **Biomes**: Savanna, Nama-Karoo and Succulent Karoo.

General A very palatable species with a high leaf yield. Although the grass is hard, it is well utilized, particularly by cattle. Can be identified by its fragile, brittle culms. **Grazing value** high to very high. **Ecological status**: mostly Decreaser.

ERAGROSTIS CILIARIS

Woolly Love Grass
Wollerige-eragrostis

A tufted annual with culms often geniculate and up to 0,6 m tall. **Inflorescence** a contracted, often interrupted, woolly panicle up to 160 mm long, consisting of groups of spikelets on short branches. Flowers throughout the year. **Spikelets** up to 4,5 mm long, purple, with a few stiff hairs. **Leaf blade** up to 5 mm wide, flat, firm, with scabrid leaf margins. **Leaf sheath** glabrous or with scattered hairs. **Ligule** a ring of short hairs.

Habitat   In bushveld in disturbed areas such as cultivated and uncultivated lands, in overgrazed and trampled veld, and often around human dwellings. Prefers moist sandy soils. (Rarely found on heavy soils.) **Biome**: Savanna.

General   A relatively palatable species with probably an average grazing value. A weed in disturbed areas, but easily controlled mechanically. The hairy spikelets of *Eragrostis ciliaris* can be confused with those of *Eragrostis viscosa*, *Eragrostis tenella* and *Eragrostis arenicola*, but *Eragrostis viscosa* and *Eragrostis tenella* have more open inflorescences, and the spikelets of *Eragrostis arenicola* are not as woolly as those of *Eragrostis ciliaris*, and the plant is often shorter (up to 350 mm tall).

HELICTOTRICHON TURGIDULUM

Small Oat Grass
Kleinhawergras

A tufted perennial, with culms up to 1,0 m tall. **Inflorescence** a dense, contracted panicle, up to 300 mm long. Flowers from October to April. **Spikelets** up to 12 mm long, with a bent awn emerging from the middle of the glabrous lemmas, the lemma lobes are up to 5 mm long. **Leaf blade** up to 6 mm wide and flattened. **Ligule** varies from a membrane to a ring of hairs.

Habitat Usually grows on slopes in higher rainfall regions and in wet areas such as vleis in the lower rainfall parts of its distribution. Sometimes in thick stands along roadsides. Grows in most kinds of soil. **Biomes**: Grassland, Savanna and Fynbos.

General A reasonably valuable grass, which remains green well into winter. *Helictotrichon turgidulum* is closely related to *Helictotrichon dodii*, *Helictotrichon longifolium* and *Helictotrichon natalense*. However, *Helictotrichon dodii* has longer lemma lobes (up to 8 mm long), *Helictotrichon longifolium* has filiform leaves and *Helictotrichon natalense* has smaller spikelets (up to 9 mm long). **Grazing value** variable. **Ecological status**: mostly Decreaser.

* LAGURUS OVATUS

Hare's Tail
Haasstert

A tufted annual with culms up to 0,6 mm high. **Inflorescence** up to 60 mm long, an egg-shaped, spike-like panicle with dense, soft, white hairs. Flowers from October to December. **Spikelets** up to 10 mm long, the glumes covered with fine white hairs. Each glume and lemma ends in a fine awn up to 25 mm long. **Leaf blade** up to 10 mm wide, flattened, thick and velvety due to an abundance of short, soft hairs. **Leaf sheath** round and membranous. **Ligule** a pubescent white membrane.

Habitat Common as a weed on disturbed sandy soils, particularly along roads and in gardens. Often found under trees and shrubs. **Biomes**: Savanna and Fynbos.

General Utilized by stock, but has low grazing value due to low leaf production. The species originates from the Mediterranean areas, but is now widespread in regions with similar climate. It is an attractive grass with ornamental value. The panicles are dried and sometimes dyed for use in flower arrangements and other decorations.

MELICA DECUMBENS

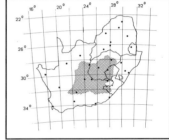

Staggers Grass
Dronkgras

A tufted perennial with culms up to 0,5 m tall and usually unbranched. **Inflorescence** a single unilateral, hairy raceme, up to 120 mm long. Flowers from October to April. **Spikelets** up to 15 mm long, glumes papery and smooth, backs of lemmas distinctly hairy. **Leaf blade** up to 3,5 mm wide, erect, glabrous, very rough and usually rolled. **Ligule** a toothed, pellicular membrane.

Habitat Usually in the shade of trees and shrubs or between rocks, often on slopes, but also on plains and sometimes along roadsides. **Biomes**: Nama-Karoo and Grassland.

General An evergreen grass with scabrid leaves. When grazed in large amounts it is toxic for cattle, horses and donkeys and sometimes sheep (see common name). Is confused with *Melica racemosa* and *Ehrharta villosa*. *Melica racemosa* has smaller spikelets (up to 9 mm long), the backs of the lemmas are glabrous, and the margins of the lemmas are hairy. *Ehrharta villosa* is a taller plant (up to 1,5 m) and the leaf blade is also rolled, but not rough. **Grazing value** very low. **Ecological status**: variable but mostly Increaser IIc.

MELICA RACEMOSA

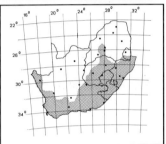

Fluffy Grass
Flossiegras

A tufted perennial with branched or unbranched culms, up to 0,5 m tall. **Inflorescence** a narrow panicle or raceme, up to 200 mm long. Flowers from September to April. **Spikelets** up to 10 mm long, lemmas with hairy margins and glabrous backs. **Leaf blade** up to 5 mm wide, flattened or rolled, glabrous and often scabrid. **Ligule** membranous.

Habitat Usually found on slopes between rocks, or in light shade. Often grows along the edges of densely wooded areas and dune forests. **Biomes:** Fynbos, Grassland and Nama-Karoo.

General Rarely an important component of natural veld. Often confused with *Melica decumbens* which has larger spikelets (up to 15 mm long) and the backs of the lemmas are covered with long hairs.

SETARIA LINDENBERGIANA

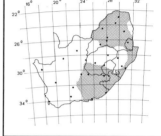

Mountain Bristle Grass
Randjiesgras

A tufted perennial with a short rhizome and culms up to 1,2 m long, usually branched. **Inflorescence** a contracted, dense to slightly open panicle, up to 200 mm long. Flowers from October to May. **Spikelets** up to 3,5 mm long, acute, with a few solitary bristles in between. **Leaf blade** up to 7 mm wide, linear, delicately pleated and scabrid, with fine to dense hairs. **Leaf sheath** compressed, keeled, usually hairy. **Ligula** a ring of hairs.

Habitat   Usually against rocky or stony slopes and in ravines. Often grows in rock crevices and in the shade of trees. Also found in open bushveld and forests. **Biomes**: Savanne, Grassland and Fynbos.

General   A palatable grass which makes good hay. Reasonably resistant to drought, but not to frost. Can be confused with *Setaria megaphylla* and *Setaria plicatilis*. *Setaria megaphylla* is a robust plant, up to 3 m tall, with pleated, lanceolate leaves 10—110 mm wide. *Setaria plicatilis* is an open, tufted grass, with lanceolate, pleated leaves up to 35 mm wide.

SETARIA VERTICILLATA

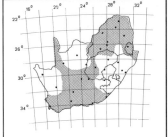

Bur Bristle Grass
Klitsgras

A tufted annual with culms up to 1,0 m tall, sometimes rooting from the lower nodes. **Inflorescence** a dense, spike-like, erect or drooping panicle up to 150 mm long. Flowers from December to April. **Spikelets** up to 2,5 mm long, glabrous, sometimes purplish. Bristles subtending the spikelets and furnished with backward pointing hooks. **Leaf blade** up to 20 mm wide, generally with a velvety surface and flattened. **Leaf sheath** compressed. **Ligule** a ring of short hairs.

Habitat Generally on nitrogen-rich disturbed soils. Often found under trees and in other shady places. Also as a weed in disturbed areas such as gardens and cultivated lands. **Biomes**: Grassland, Savanna, Nama-Karoo and Fynbos.

General A palatable grass with a variable leaf production, well grazed during winter and summer. Also suitable for hay-making. A serious problem with sheep and angora goats, where the inflorescence clings to the animals' coats, thus lowering the quality of the wool and hair. Can be a troublesome weed in gardens, vineyards, cultivated lands and other disturbed areas. **Grazing value** low. **Ecological status**: Increaser IIc.

SPOROBOLUS AFRICANUS

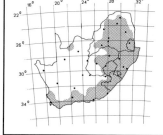

Ratstail Dropseed
Taaipol

A tufted perennial with erect or oblique culms up to 1,5 m tall. **Inflorescence** a dense contracted panicle with relatively short branches. Flowers from October to April. **Spikelets** up to 2,8 mm long, glumes unequal and upper glume sharply pointed. **Leaf blade** up to 4 mm wide, glabrous and often shiny. **Leaf sheath** glabrous and slightly compressed. **Ligule** inconspicuous.

Habitat   Usually grows in disturbed areas such as overgrazed veld or along roads on compacted soils. Often found near streams, rivers or watercourses. **Biomes**: Grassland, Savanna and Fynbos.

General   A palatable but tough grass with a low leaf production. A good indicator of disturbed veld. Traditionally used for medicinal purposes like application to wounds and snake-bites. *Sporobolus africanus* is closely related to *Sporobolus fimbriatus*, *Sporobolus natalensis* and *Sporobolus pyramidalis* and intermediates often occur. *Sporobolus africanus* can be distinguished from the others by its contracted, spike-like panicle. **Grazing value** low. **Ecological status**: mostly Increaser IIc.

CTENIUM CONCINNUM

Sickle Grass
Sekelgras

A tufted, perennial with unbranched culms up to 1,0 m tall. **Inflorescence** a solitary, unilateral spike, up to 170 mm long, typically sickle-shaped when young and spirally twisted like a cork-screw when mature. Flowers from December to April. **Spikelets** up to 7 mm long, unilaterally arranged with conspicuous, straight awns. **Leaf blade** up to 5 mm wide, flat or rolled and concentrated mostly around the base of the plant. **Leaf sheath** generally round. **Ligule** a short membrane.

Habitat Commonly found in open grassland and sometimes also in open bushveld. Prefers poor, dry, sandy soils but sometimes also grows on moist soils. **Biomes**: Grassland and Savanna.

General Probably has a low grazing value, owing to its hard leaves.
Some species of *Microchloa* and *Enteropogon* have similar inflorescences, but *Ctenium concinnum* differs in that the inflorescence is much thicker in proportion to the length.

DIGITARIA MONODACTYLA

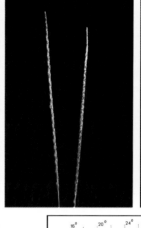

One-finger Grass
Eenvingergras

A tufted perennial, with culms up to 0,55 m tall (rarely up to 0,9 m). **Inflorescence** a solitary spike-like raceme, up to 180 mm long, unilateral and finely hairy. Flowers from November to March. **Spikelets** up to 3 mm long, in pairs, shortly pedicellate, hairy. **Leaf blade** up to 2 mm wide, often rolled, densely hairy to almost glabrous. **Leaf sheath** round and finely hairy. **Ligule** a short papery membrane.

Habitat Generally occurs in open sour grassland or in open patches in sour bushveld. Found in disturbed areas, but also in stable grassland. Grows on most soil types, but prefers sandy soil. **Biomes**: Grassland and Savanna.

General A relatively palatable species, which is well utilized in certain areas, but usually has a low leaf production. Could be confused with *Elionurus muticus*, but the latter has a spike typically curled when dry. May also be confused with *Microchloa caffra*, but the latter is a much smaller plant with curved inflorescences. **Grazing value** low to average. **Ecological status**: mostly Increaser IIb.

ENTEROPOGON MONOSTACHYUS subsp. AFRICANUS

Needle Grass
Naaldgras

A tufted perennial with culms up to 1,0 m long. **Inflorescence** a lax, solitary, unilateral spike up to 200 mm long. Flowers from November to April. **Spikelets** of 2—3 florets, with lemmas of lower floret up to 8 mm long and bearing an awn up to 8 mm long. **Leaf blade** up to 4 mm wide. **Leaf sheath:** basal sheaths laterally flattened and strongly keeled. **Ligule** a short membrane with a hairy margin.

Habitat Usually near rivers or in other low-lying areas. Grows on granite plains and sandy soil, often in the shade of trees. **Biome:** Savanna.

General A relatively unknown species. *Enteropogon monostachyus* can be distinguished from *Enteropogon macrostachyus* and *Enteropogon rupestris* by the round leaf sheaths and glabrous sheath margins of the latter two species.

HARPOCHLOA FALX

Caterpillar Grass
Ruspergras

A tufted perennial with a short rhizome and culms up to 0,9 m tall. **Inflorescence** a single unilateral spike, up to 80 mm long, which curves back in a sickle shape when mature. Flowers from September to April. **Spikelets** up to 9 mm long, flattened, woolly and arranged in two rows on the rachis. **Leaf blade** up to 4 mm wide, usually inrolled, glabrous, with a blunt tip. **Ligule** inconspicuous, with long white hairs along the margins.

Habitat Occurs on moist stony slopes on well drained stony soils. Prefers undisturbed open and mountainous grassland. **Biomes**: Grassland, Savanna and Fynbos.

General A relatively palatable grass with a high leaf production which is well utilized in a early stage by grazers. However, it becomes hard and unacceptable as pasture when mature. Often forms dense stands, particularly in high-lying mountain veld. **Grazing value** average. **Ecological status**: mostly Increaser I.

MICROCHLOA CAFFRA

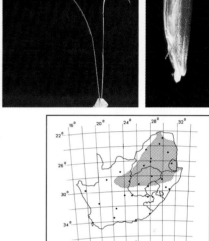

Pincushion Grass
Elsgras

A tufted perennial with unbranched culms up to 0,5 m tall. **Inflorescence** a unilateral spike up to 150 mm long which is initially straight but soon becomes curved. Flowers from October to April. **Spikelets** up to 5,5 mm long, arranged unilaterally on the rachis, glabrous, with equal glumes. **Leaf blade** up to 1,5 mm wide, rolled, curled, with inconspicuous hairs. **Leaf sheath:** basal sheaths break up into fine fibres when old, usually glabrous. **Ligule** an inconspicuous ring of short hairs.

Habitat Associated with shallow soils, stony soils and rocky outcrops, where the grass can survive on virtually only humus and very little water. Often found in disturbed and overgrazed areas and sometimes near vleis. Grows on most soil types, but prefers sandy soils. **Biomes**: Grassland, Savanna en Nama-Karoo.

General Although the palatability of the grass varies from region to region, it has a very low leaf production and is therefore of very low grazing value. However, it plays an important role in the stabilization of bare shallow soils. Although this grass is never dominant in the veld, it can increase during overgrazing. *Microchloa caffra* can easily be confused with *Microchloa kunthii*, but the latter usually has shorter spikelets. **Grazing value** very low. **Ecological status**: Increaser IIc.

STENOTAPHRUM SECUNDATUM

Coastal Buffalo Grass
Strandbuffelsgras

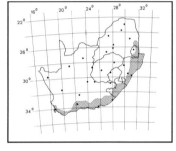

A prostrate perennial with extensive stolons and branched culms up to 350 mm tall. **Inflorescence** spike-like, up to 150 mm long, consisting of a thickened, flattened rachis and impressed spikelets. Flowers from October to May. **Spikelets** sessile, glabrous, up to 5 mm long. **Leaf blade** up to 10 mm wide, relatively short, glabrous and obtuse. **Leaf sheath** compressed, smooth and overlapping. **Ligule** a ring of short hairs.

Habitat Common in coastal regions near the sea or fresh water sources. Grows on most types of soil, with a preference for sandy soils.

General A palatable grass which is well utilized, particularly when growing in fertile soil with sufficient water. Commonly used for course lawns. An important dune and sand stabilizer in coastal regions. **Grazing value** high. **Ecological status**: Decreaser.

DIGITATE INFLORESCENCES

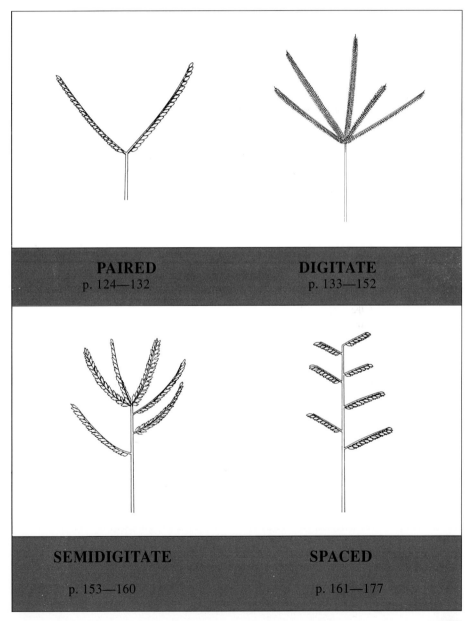

PAIRED
p. 124—132

DIGITATE
p. 133—152

SEMIDIGITATE
p. 153—160

SPACED
p. 161—177

ANDROPOGON CHINENSIS

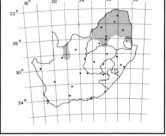

Hairy Blue Grass
Harige-blougras

A tufted perennial, with branched culms up to 1,2 m tall. **Inflorescence** consists of paired racemes (occasionally three) up to 75 mm long, gathered in loose false panicles and with inconspicuous long soft hairs. Flowers from December to June. **Spikelets** up to 7 mm long, occur in pairs: one spikelet sessile with 2 awns (one straight and 10 mm long, the other curved and up to 35 mm long), the other spikelet is pedicellate, with two straight awns up to 6 mm long. **Leaf blade** up to 8 mm wide, usually glabrous, except for a few fine hairs sometimes near the ligule. **Leaf sheath** round, glabrous and with auricles. **Ligule** membranous with long stiff hairs on the sides.

Habitat   Generally in bushveld, but sometimes also in open grassveld. Usually found on stony slopes. Prefers stony, sandy soils, but also grows on loamy soils. **Biomes**: Savanna, Grassland and Nama-Karoo.

General   A valuable pasture grass which is reasonably well utilized, especially in the young stage. The entire plant has a blue-green to grey colour, with a shiny or waxy appearance. Sometimes confused with *Andropogon schirensis* which has only one pair of racemes per culm, and with *Diheteropogon amplectens*, which has rounded leaf blade bases.

ANDROPOGON SCHIRENSIS

Stab Grass
Tweevingergras

A tufted perennial with unbranched culms, up tot 1,2 m tall. **Inflorescence** a pair of racemes, up to 150 mm long, borne terminally on the culm. Flowers from December to April. **Spikelets** paired, with one member sessile, up to 7 mm long, glabrous and with an awn up to 35 mm long, and the other awnless with a hairy pedicel. **Leaf blade** up to 14 mm wide, glabrous or hairy, flattened.
Ligule a membrane.

Habitat Occurs in grassland or open bushveld, particularly on stony slopes. Usually grows in well drained stony soils, but sometimes also in moist areas. **Biomes**: Grassland and Savanna.

General A grass of medium palatability which decreases as the plant matures. The entire plant often has a reddish brown colour. *Andropogon schirensis* could be confused with *Andropogon chinensis* and *Diheteropogon amplectens*. *Andropogon chinensis* has branched culms and *Diheteropogon amplectens* has rounded leaf bases. **Grazing value** low to average. **Ecological status**: Increaser I or Decreaser.

125

DACTYLOCTENIUM AUSTRALE

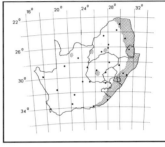

L.M. Grass
L.M.-gras

A creeping perennial with stolons, culms up to 0,6 m tall (often shorter). **Inflorescence** consists of 2 or 3 (occasionally 4) unilateral spikes, up to 50 mm long, arranged digitately at the tip of the culm. Flowers from January to May. **Spikelets** up to 4 mm long, flattened, with a short awn up to 0,7 mm long. **Leaf blade** up to 4,5 mm wide, with evenly spaced hairs on the leaf margins. **Leaf sheath** slightly compressed. **Ligule** a short membrane.

Habitat Occurs naturally in warm subtropical and tropical regions, generally in sandy soils, and often in shade. **Biome**: Savanna.

General A palatable pasture grass, well utilized by stock. Also a popular evergreen lawn in warm areas, which grows well in shade. Used for stabilising inland and coastal quicksand. Often confused with the closely related *Dactyloctenium geminatum*, but the latter is a taller and more robust plant than *Dactyloctenium australe*. **Grazing value** average to high. **Ecological status**: mostly Decreaser.

DACTYLOCTENIUM GEMINATUM

Sign Grass
Wysergras

A creeping perennial with stolons, culms up to 0,7 m tall. **Inflorescence** consists of 2 or 3 (rarely more) unilateral spikes up to 70 mm long, arranged digitately at the tip of the culm. Flowers from December to March. **Spikelets** up to 5 mm long, flattened, lemma acute. **Leaf blade** up to 6 mm wide.

Habitat   Occurs near pans and rivers or in other low-lying areas on brackish or sandy soils in coastal regions. Sometimes found along roads where water accumulates. **Biome**: Savanna

General   Probably palatable, but with a relatively low leaf production. Plays an important role in erosion control on brackish and sandy soils. Often confused with *Dactyloctenium australe*, but can be distinguished by *Dactyloctenium geminatum* being a taller and more robust plant.

DIGITARIA EYLESII

Eyle's Finger Grass
Swartsaadtweevingergras

A tufted perennial with a creeping rhizome and unbranched culms, up to 1,0 m tall. **Inflorescence** consists of 1—3 (usually 2) thin racemes, one of them usually longer than the other, the length varying from 100—200 mm. Flowers from January to April. **Spikelets** up to 3 mm long, glabrous, usually in groups of three, with unequal pedicels. **Leaf blade** up to 4 mm wide, flattened, erect and glabrous or inconspicuously hairy. **Leaf sheath** round. **Ligule** a short membrane.

Habitat   Grows in wet areas such as vleis and riverbanks. **Biomes**: Savanna and Grassland.

General   An insignificant pasture grass, because it is rarely an important component of grassland, except in some vleis, where it may form thick stands. In certain regions it plays an important role in protecting riverbanks against erosion. *Digitaria eylesii* may sometimes be confused with *Digitaria diagonalis*, but the inflorescence of the latter has much more racemes, and it does not have a creeping rhizome.

DIHETEROPOGON AMPLECTENS

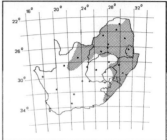

Broad-leaved Bluestem
Breëblaarblougras

A tufted perennial with erect culms 0,6—1,5 m tall. **Inflorescence** consists of two terminal racemes 90 mm long. Flowers from January to March. **Spikelets** of two types: one 6—8 mm long, sessile and with an awn, the other slightly longer, pedicellate and without an awn. **Leaf blade** up to 30 mm wide at the base and glabrous or hairy. The leaf base is typically wide and curved around the culm. **Leaf sheath** rounded. **Ligule** a short inconspicuous membrane.

Habitat   Occurs commonly in high-lying sour grassland, but also in bushveld. Grows on various soils but prefers stony soil. Often found on rocky hills. **Biomes**: Grassland and Savanna.

General   A relatively palatable species, well utilized by stock, especially early in the season. Unfortunately the grass is not very leafy and the grazing value is therefore impaired. *Diheteropogon amplectens* could possibly be confused with *Cymbopogon excavatus* which occurs in the same habitat and shows similarities in the leaves. The leaves of the latter have a strong turpentine smell and taste. Preference species for waterbuck. **Grazing value** average to high. **Ecological status**: Decreaser.

DIHETEROPOGON FILIFOLIUS

Thread-leaved Bluestem
Smalblaarblougras

A tufted perennial with unbranched culms up to 0,6 m tall. **Inflorescence** consists of two racemes up to 80 mm long, arranged terminally on the culms. Flowers from October to April. **Spikelets** in pairs, one member sessile, up to 8 mm long, with a bent awn up to 55 mm long, and the other one pedicellate, up to 17 mm long, awnless. **Leaf blade** thread-like or up to 3,5 mm wide and glabrous or hairy at the base. **Leaf sheath:** basal sheaths tear up into fibres. **Ligule** an inconspicuous ring of hairs.

Habitat Common on slopes in open mountain sourveld. However, it also grows on slopes in bushveld areas and sometimes in coastal regions, usually on sandy and loamy soils. **Biomes**: Grassland and Savanna.

General It is regarded as a hard and unpalatable grass, which is very poorly utilized when mature. *Diheteropogon filifolius* is sometimes confused with the closely related *Andropogon schirensis* or with *Diheteropogon amplectens*, but the latter two species both have flattened leaves as opposed to the thread-like leaves of *Diheteropogon filifolius*. **Grazing value** very low. **Ecological status**: mostly Increaser IIb.

PASPALUM DISTICHUM

Couch Paspalum
Kweekpaspalum

A water-loving, creeping perennial with stolons, rhizomes and rooting from the lower nodes, culms up to 300 mm long. **Inflorescence** generally consists of two racemes each up to 70 mm long, which often bend outwards at maturity. Flowers from November to May. **Spikelets** up to 3,5 mm long, arranged in two rows, glabrous, acute, (one glume usually a small triangular scale). **Leaf blade** up to 8 mm wide, flattened, glabrous, except for hairs near the ligule. **Ligule** a membrane up to 3 mm long.

Habitat Occurs mainly in or near rivers, vleis, pans, canals or other wet areas. Also found in cultivated lands in moderate- to high-rainfall regions. Grows on most types of soil, including sand and black turf.
Biomes: Grassland, Savanna, Nama-Karoo and Fynbos.

General A palatable species and particularly suitable grazing for sheep. A troublesome weed, difficult to control, in cultivated lands and canals. Confused with *Paspalum notatum* and *Paspalum vaginatum*. The spikelets of *Paspalum notatum* are more rounded and glossy than the sharp-pointed spikelets of *Paspalum distichum*, while *Paspalum vaginatum* generally occurs at the coast, and a lower glume is usually absent.

* PASPALUM NOTATUM

Bahia Grass
Bahiagras

A perennial with a strong, creeping rhizome and culms up to 0,7 m tall. **Inflorescence** consists of 2 (rarely 3) racemes up to 70 mm long, arranged digitately at the tip of the culm. Flowers from December to April. **Spikelets** up to 4 mm long, glabrous, with dark hues due to exerted purple male and female floral parts. **Leaf blade** up to 10 mm wide and generally sparsely hairy. **Ligule** an inconspicuous membrane.

Habitat   Common in moist disturbed areas of the higher rainfall regions, e.g. along roadsides, in gardens and lawns. Prefers well drained, sandy to clayey soils. **Biomes**: Grassland and Fynbos.

General   Indigenous to South America. Originally imported as artificial pasture, but now widespread in the tropical and subtropical parts of the country. Selected cultivars such as 'Paraguay' and 'Pensacola' are planted as pasture, particularly for sheep. Also used for maintaining erosion control works. **Grazing value** high. **Ecological status**: Decreaser.

ALLOTEROPSIS SEMIALATA subsp. ECKLONIANA

Black-seed Grass
Donkersaadgras

A tufted perennial, generally with a short rhizome and unbranched culms up to 1,3 m tall. **Inflorescence** consists of 2—6 racemes, up to 150 mm long, arranged digitately at the tip of the culm. Flowers from September to March. **Spikelets** up to 8 mm long (including the awn), relatively large, with a short awn. **Leaf blade** up to 12 mm wide, flattened, relatively thick with velvety hairs. **Leaf sheath:** basal sheaths distinctly ribbed and with dense white hairs. **Ligule** membranous, with a ring of short hairs.

Habitat Common in open sour grassland and in open patches in sour bushveld. Found mostly in stable veld or in protected fenced-off areas. Usually on stony acid soils. **Biomes**: Grassland and Savanna.

General An unpalatable species which is reasonably well utilized in the young stage. Thereafter the leaves become hard and unacceptable to grazers. Two subspecies are distinguished, namely *Alloteropsis semialata* subsp. *eckloniana* and *Alloteropsis semialata* subsp. *semialata*. They can be distinguished from one another by the broad, flattened, densely hairy leaves of subsp. *eckloniana* and the leaves of subsp. *semialata* which are generally narrow and sparsely hairy. **Grazing value** low to average. **Ecological status**: Increaser I.

ANDROPOGON APPENDICULATUS

Vlei Bluestem
Vleiblougras

A tufted perennial with culms up to 1,3 m tall. **Inflorescence** consists of 4—20 racemes up to 150 mm long, arranged semidigitately on a short axis. Flowers from October to April. **Spikelets** up to 7 mm long, paired, one is sessile, with an awn up to 15 mm long, and the other pedicellate, with a short awn up to 2 mm long, the pedicel with white or yellow hairs. **Leaf blade** up to 6 mm wide and folded at the base, glabrous or finely hairy near the ligule, most leaves borne basally. **Leaf sheath:** basal sheaths compressed, keeled, glabrous, glossy.

Habitat Occurs in wet areas such as vleis, stream banks and on other low-lying ground. Widespread, but more prevalent in the higher rainfall areas. **Biomes**: Grassland, Savanna and Fynbos.

General A palatable grass, well utilized by grazers. Can form dense stands in low-lying wet areas. Can withstand heavy grazing. Certain individual plants with particularly hairy racemes could be confused with *Andropogon huillensis*. However, *Andropogon huillensis* usually has 5—7 flowering branches per culm compared with the 1 or 2 flowering branches of *Andropogon appendiculatus*. **Grazing value** high. **Ecological status**: Decreaser.

* CHLORIS GAYANA

Rhodes Grass
Rhodesgras

A tufted perennial or weak perennial spreading by means of stolons. Culms up to 1,5 m tall (rarely taller). **Inflorescence** consists of up to 20 racemes arranged digitately in one or two whorls on the axis. Flowers from November to May. **Spikelets** up to 5 mm long, flattened, bearing two awns up to 10 mm long. **Leaf blade** flattened and smooth. **Leaf sheath** compressed and keeled. **Ligule** a membrane with a hairy margin.

Habitat Improved strains grow well on all well-drained soils. Under natural conditions, the grass prefers moist sites such as vleis and riverbanks. **Biomes**: Savanna, Grassland and Fynbos.

General Naturalized from India. A palatable to highly palatable forage grass with a high leaf production and average nutritive value. Selected strains such as 'Katambora' and 'Giant Rhodes' are planted as pasture and used to maintain erosion-control works in many parts of the world. *Chloris gayana* is sometimes confused with *Eustachys paspaloides*. However, *Chloris gayana* has more racemes which are generally more flexible. It can furthermore be distinguished by the spikelets bearing longer awns. **Grazing value** high. **Ecological status**: Increaser IIb under natural conditions.

CHLORIS MOSSAMBICENSIS

Perennial spiderweb Grass
Meerjarige Spinnerakgras

A tufted perennial with rhizomes and stolons, culms up to 0,5 m tall. **Inflorescence** digitate, consisting of 4 or 5 spikes up to 80 mm long. Flowers from October to April. **Spikelets** up to 4 mm long, with two awns up to 11 mm long. **Leaf blade** up to 6 mm wide. **Leaf sheath:** basal sheaths strongly keeled.

Habitat Generally in wet areas such as riverbanks or seasonal pans. Grows on clay and waterlogged soil. **Biomes:** Savanna and Grassland.

General Has a low leaf production and is therefore of little value for grazing. Can be distinguished from *Chloris pycnothrix* and *Chloris virgata* by the fact that these are both annuals without rhizomes or stolons. *Chloris gayana* is a larger plant (up to 1,5 m tall) with a higher leaf production than *Chloris mossambicensis*.

CHLORIS PYCNOTHRIX

Spiderweb Grass
Spinnerakgras

A tufted annual to weak perennial with culms up to 0,6 m tall, often geniculate, rooting at the lower nodes. **Inflorescence** consists of 4—9 spikes, up to 100 mm long, arranged digitately or semidigitately on the culm. Flowers from September to May. **Spikelets** up to 3 mm long, arranged unilaterally in two rows, with conspicuously long awns up to 27 mm. **Leaf blade** up to 5 mm wide, with a distinct midrib and a typically blunt tip. **Leaf sheath** strongly compressed. **Ligule** a membrane with a hairy margin.

Habitat Common in disturbed areas such as cultivated lands, gardens and roadsides. Grows on all soil types, with a preference for red and stony soils. **Biomes**: Grassland and Savanna.

General Rarely in natural veld, and with a low leaf production. A common weed, which may become a problem, particularly in gardens and lawns. Plays an important role in soil erosion control, because it often colonizes and thus stabilizes disturbed soils. **Grazing value** very low to low. **Ecological status**: Increaser IIc.

CHLORIS VIRGATA

Feathered Chloris
Witpluim-chloris

A tufted annual to weak perennial with culms up to 0,8 m tall, and rooting at the lower nodes. **Inflorescence** consists of 4—15 feathery spikes, arranged digitately in one whorl around the apex of the culms. Flowers from December to June. **Spikelets** up to 3,5 mm long, hairy and flattened, sometimes black when ripe, with two awns. **Leaf blade** up to 6 mm wide, flattened with a distinctive midrib, rough on the margins, green to blue-green. **Leaf sheath** compressed and keeled. **Ligule**: an inconspicuous short membrane.

Habitat Generally in gardens, uncultivated lands and other disturbed sites on all soil types, particularly where water accumulates after rain. In the drier parts of the country the grass occurs mainly on heavier moist soils adjoining pans, vleis etc. **Biomes**: Savanna, Grassland, Nama-Karoo and Succulent Karoo.

General A fast-growing palatable grass, well utilized by livestock and game, but with a limited leaf production. A typical weed in the maize producing parts of the country and it is inclined to increase during high-rainfall seasons. This grass grows on bare compacted soils where few other grasses survive and it is one of the first grasses to start the succession process. **Grazing value** generally low. **Ecological status:** Increaser IIc.

CYNODON DACTYLON

Couch Grass
Kweek

A creeping perennial forming a thick mat by means of stolons and rhizomes. Culms up to 450 mm long, rooting at lower nodes. **Inflorescence** digitate, consisting of 3—7 spikes up to 60 mm long, arranged terminally on the axis. Flowers from September to March. **Spikelets** up to 2,5 mm long, sessile and without an awn. **Leaf blade** up to 4 mm wide, flattened with a sharp tip, hairy or glabrous. **Leaf sheath** round and glabrous except for hairs sometimes occurring on the lower leaf sheaths. **Ligule** an inconspicuous ring of hairs or a short membrane.

Habitat Occurs on almost all soil types, but has a preference for soils with a high nitrogen content. Occurs in disturbed areas such as roadsides, gardens and uncultivated lands. Often found in moist sites along rivers and below the embankments of dams. **Biomes**: Grassland, Savanna, Nama-Karoo and Fynbos.

General It is a relatively good pasture grass and can withstand intensive grazing. In areas with mild winters, it remains green until late in the season. The grass plays an important role in natural soil erosion control because of its lawn-like habit and is often established in waterways. It is important as a cultivated pasture and well known cultivars such as 'NK37' and 'Bermuda' have been selected. It is difficult to control mechanically and it constitutes a weed problem in cultivated lands. **Grazing value** average under natural conditions. **Ecological status**: Increaser IIb.

* CYNODON NLEMFUENSIS

Star Grass
Stergras

A perennial with strong stolons, up to 1,5 m and longer, culms up to 0,6 m (rarely up to 1,2 m) tall. **Inflorescence** consisting of 7—12 slender spikes, up to 130 mm long, arranged in one (occasionally two) whorl on a short axis. Flowers from January to March. **Spikelets** up to 3 mm long, awnless and purple, lemmas with dense fine hairs. **Leaf blade** up to 6 mm wide, rough and flat or folded. **Leaf sheath** glabrous and keeled. **Ligule** an inconspicuous membrane.

Habitat Occurs in open patches in open and dense bushveld. Prefers disturbed areas such as roadsides, uncultivated lands and cattle kraals. Also found in moist areas near rivers and streams. Grows on most soil types. **Biome**: Savanna.

General Naturalized from Kenya. A palatable, aggressive and leafy grass which is frequently planted for pasture and used for hay. Selected cultivars such as 'De Hoek' and 'Estcourt' are successfully planted in warm regions with an annual rainfall of 650 mm or more. Planted by means of runners, because of a shortage of good quality seed. Sometimes a weed in orchards and fields, which is difficult to eradicate mechanically. The other well known and closely related Star Grass is *Cynodon aethiopicus*, which is a larger and more robust plant, with the spikes arranged in up to 5 whorls on the axis, and the lemma keel usually glabrous. **Grazing value** high.

DACTYLOCTENIUM AEGYPTIUM

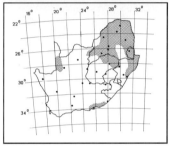

Common Crowfoot
Hoenderspoor

A tufted annual with culms up to 0,6 m tall, rooting at the lower nodes. **Inflorescence** consists of 3—8 unilateral spikes, arranged digitately at the apex of the culm. Flowers from January to April. **Spikelets** up to 4,5 mm long, flattened, sessile and bearing an awn up to 2,5 mm long at the tip of the upper glume. **Leaf blade** up to 8 mm wide, flattened with a distinct midrib and spaced hairs on the leaf margins. **Leaf sheath** keeled and slightly compressed. **Ligule** a membrane, with an uneven margin.

Habitat Occurs in disturbed areas near water. A weed in disturbed areas such as cultivated lands, gardens and roadsides. Found on all soil types. **Biomes**: Nama-Karoo, Savanna and Grassland.

General A palatable pasture and fodder grass, well utilized by grazers, particularly in the young stage. However, rarely an important component of natural veld. In times of famine, the 'seed' has been used as food in India and Africa, but it is apparently not very palatable. *Dactyloctenium aegyptium* may be confused with *Eleusine coracana* subsp. *africana* but can be distinguished by absence of an awn in the spikelets of the latter. **Grazing value** average. **Ecological status**: Increaser IIc.

DACTYLOCTENIUM GIGANTEUM

Giant Crowfoot
Reuse Hoenderspoor

A tufted annual with branched culms up to 1,0 m tall, often roots at the lower nodes. **Inflorescence** 3—9 unilateral spikes up to 110 mm long, arranged in a whorl at the tip of the culm. Flowers from November to May. **Spikelets** up to 6 mm long, strongly flattened, bearing a short awn up to 2 mm long. **Leaf blade** up to 12 mm wide, with a distinct midrib and sparse hairs on the leaf margins. **Leaf sheath** keeled. **Ligule** a membrane with a margin of short hairs.

Habitat A sandveld species, may form dense stands during seasons of good rainfall. Occurs particularly in disturbed places such as roadsides, uncultivated lands and trampled areas. Can also be a weed in cultivated lands. Sometimes found under trees or along riverbanks near water. **Biomes**: Savanna and Grassland.

General A palatable pasture grass, with a high yield, well utilized by livestock and game, particularly in the young stage. Makes good hay if cut in the early flowering stage. Sometimes confused with *Eustachys paspaloides* and *Chloris gayana,* but *Eustachys paspaloides* is an erect perennial grass with flattened, compressed basal leaf sheaths; and in *Chloris gayana* the racemes of the inflorescence are more slender and less firm than those of *Dactyloctenium giganteum*. Preferred by gemsbok.

DIGITARIA ARGYROGRAPTA

Silver Finger Grass
Silvervingergras

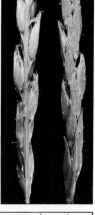

A tufted perennial with a rhizome and culms up to 0,6 m tall, branching from the lower nodes. **Inflorescence** consists of two or three upright racemes, up to 100 mm long, borne closely together, racemes rarely of the same length. Flowers from November to March. **Spikelets** up to 3 mm long, with sharp tips and silky white hairs. **Leaf blade** up to 3 mm wide, flattened or folded, glabrous or sparsely hairy, keeled. **Leaf sheath**: basal sheaths hairy, particularly when young. **Ligule** a papery membrane.

Habitat   Occurs in open grassland or in bushveld on most types of soil, but generally on stony or heavier soils. Adapts to a wide range of habitats. **Biomes**: Grassland, Savanna and Nama-Karoo.

General   A palatable grass with a reasonable leaf production, which is well utilized during the growing season. Drought-resistant. Preferred by blue wildebeest. **Grazing value** high. **Ecological status**: mostly Decreaser.

DIGITARIA BRAZZAE

Brown Finger Grass
Bruinvingergras

A tufted perennial with unbranched culms up to 1,1 m long, with hairy nodes. **Inflorescence** up to 200 mm long, consisting of 2—4 spike-like racemes arranged digitately on a short axis. Flowers from December to April. **Spikelets** up to 3,2 mm long, in groups of 2—4 and conspicuously hairy. **Leaf blade** up to 4 mm wide, basal leaves flat, with velvety hairs. **Leaf sheath:** basal sheaths densely hairy. **Ligule** a membrane with a hairy margin.

Habitat Generally in sour grassland on sandy soils, often against stony slopes and sometimes in open bushveld. **Biomes**: Grassland and Savanna.

General A relatively unknown species, which probably has a average grazing value. Rarely an important component of natural veld. *Digitaria brazzae* could be confused with *Digitaria tricholaenoides*, but the latter is a shorter plant (up to 0,55 m tall) with a horizontal rhizome.

DIGITARIA ERIANTHA

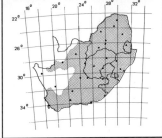

Finger Grass
Vingergras

A tufted perennial with a compact rhizome and sometimes stolons, culms up to 1,0 m tall (seldom up to 1,4 m). **Inflorescence** 3—15 racemes, up to 200 mm long, arranged in one or two whorls at the tip of the culm. Flowers from January to April. **Spikelets** up to 4 mm long, hairy. The lower glume is a membranous scale sheath. **Leaf blade** up to 14 mm wide, hairy and flat or folded. **Leaf sheath:** basal sheaths often densely hairy. **Ligule** a membrane up to 5 mm long, with a hairy collar.

Habitat Widely adapted to a variety of habitats. Occurs in eastern high-rainfall areas, especially on damp soil along vleis and rivers in tall grassland. On the highveld and in the drier western parts the grass grows in sandy and stony soil on plains and hills. **Biomes:** Savanna, Grassland, Fynbos and Nama-Karoo.

General A highly digestible and palatable pasture grass, with a high leaf production. Well utilized by grazers. Under natural conditions it can be an indicator of good veld conditions. A well known pasture grass (Smuts Finger Grass) which provides good quality hay and grazing. Often sown into veld where the conditions are poor. A very variable grass with many forms. Preferred by impala and roan. **Grazing value** mostly very high. **Ecological status**: Decreaser.

DIGITARIA MAITLANDII

Drakensberg Finger Grass
Drakensberg-vingergras

A tufted perennial with unbranched culms up to 400 mm long. **Inflorescence** consists of up to seven racemes up to 80 mm long, arranged semidigitately on a short primary axis. Flowers from November to May. **Spikelets** up to 2 mm long, occur in groups of three, glabrous or sparsely hairy. **Leaf blade** up to 4 mm wide, flattened, and usually densely hairy.

Habitat   Generally on slopes in mountainous grassland. Often grows on loamy soils. **Biome**: Grassland.

General   A relatively unknown species which is sometimes confused with other 'dark-seeded' species of *Digitaria* such as *Digitaria diagonalis* var. *diagonalis*, *Digitaria eylesii* and *Digitaria ternata*, but can be distinguished as follows: *Digitaria diagonalis* var. *diagonalis* is a robust plant with culms up to 1,5 m tall; *Digitaria eylesii* has a rhizome and usually only two racemes per inflorescence; *Digitaria ternata* is an annual and its spikelets are usually densely hairy.

* DIGITARIA SANGUINALIS

Crab Finger Grass
Kruisvingergras

A tufted annual with culms up to 0,6 m tall and often branched, rooting at the lower nodes. **Inflorescence** consists of 3—12 racemes up to 150 mm long, arranged digitately or semidigitately on the primary axis. Flowers from November to May. **Spikelets** up to 3,5 mm long, occurring in pairs of which one member is pedicellate and the other is sessile or shortly pedicellate. **Leaf blade** up to 10 mm wide, usually hairy, flattened and thin. **Ligule** a membrane, up to 2 mm long.

Habitat   Occurs mainly in disturbed areas, such as gardens and cultivated fields. Prefers regions with a mild climate. **Biomes**: Savanne, Grassland, Fynbos, Nama-Karoo and Succulent Karoo.

General   An important weed in gardens and fields worldwide. Originally from Europe. Often confused with *Digitaria ciliaris*, but the leaf blade of *Digitaria ciliaris* is usually glabrous and the ligule more conspicuous. **Grazing value** low. **Ecological status**: Increaser IIc.

DIGITARIA TERNATA

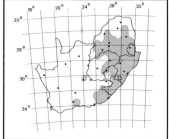

Black-seed Finger Grass
Swartsaadvingergras

A tufted annual with culms up to 0,6 m tall. **Inflorescence** consists of 2—7 racemes, arranged semidigitately on a short primary axis with fine white hairs to just below the inflorescence. Flowers from December to May. **Spikelets** hairy, awnless, with glumes very unequal. 'Seeds' dark purple or black. **Leaf blade** glabrous, flattened, relatively short, basal leaves often red. **Leaf sheath** glabrous and rounded or keeled. **Ligule** a firm membrane.

Habitat   Usually on uncultivated lands, along roadsides and in other disturbed sites, particularly on wet compacted soils. **Biomes**: Savanna and Grassland.

General   A palatable grass, probably with a low grazing value as a result of its relatively low leaf production. Seen as a weed, particularly in gardens.

DIGITARIA TRICHOLAENOIDES

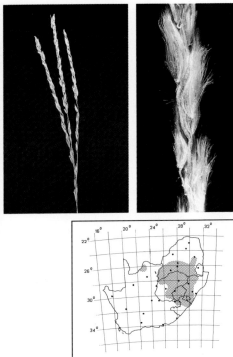

Purple Finger Grass
Persvingergras

A tufted perennial with a strong rhizome and culms up to 0,55 m tall. **Inflorescence** consists of 2—7 racemes, up to 130 mm long, arranged digitately or semidigitately on the primary axis. Flowers from November to March. **Spikelets** up to 5 mm long, in groups of 2—5 and covered with silver or purple hairs. **Leaf blade** up to 7 mm wide, flat and hairy or glabrous. **Leaf sheath** densely hairy, with basal sheaths overlapping and covering the rhizome. **Ligule** a membrane.

Habitat Generally in open sour grassland and sometimes in sour bushveld areas, mostly in undisturbed veld. Often found on stony slopes. Prefers stony soils. **Biomes**: Grassland and Savanna.

General A palatable to very palatable grass, well utilized by livestock. It is, however, sensitive to selective overgrazing, and therefore often prevented from reaching the flowering stage when it is heavily grazed. Its mat-forming growth form enables it to withstand trampling reasonably well, and it reacts positively when the veld is rested. *Digitaria tricholaenoides* may be confused with *Digitaria brazzae*, but it differs in that the latter does not have a rhizome and the plant is taller (up to 1,1 m). **Grazing value** high. **Ecological status**: Decreaser.

DIGITARIA VELUTINA

Flaccid Finger Grass
Slapvingergras

A tufted annual with branched culms up to 0,8 m tall, often with roots at the lower nodes and hairs below the inflorescence. **Inflorescence** consists of 6—15 lax racemes, up to 100 mm long, arranged digitately or semidigitately on the primary axis. Flowers from December to May. **Spikelets** up to 2 mm long, arranged unilaterally and with lower glume reduced. **Leaf blade** up to 10 mm wide, flattened, thin and lax. **Leaf sheath** keeled and hairy. **Ligule** membranous.

Habitat Generally in disturbed areas such as gardens, uncultivated lands and roadsides, often in moist places and in shade. Prefers sandy soils. **Biomes**: Savanna and Grassland.

General Probably of little grazing value, owing to its low leaf production. Sometimes a weed in cultivated fields, but is easily controlled mechanically.

EULALIA VILLOSA

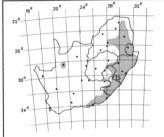

Golden Velvet Grass
Geelfluweelgras

A tufted perennial with unbranched, generally hairy culms and up to 1,4 m tall. **Inflorescence** a group of 2—7 hairy racemes, each up to 200 mm long, arranged digitately around a short primary axis. Flowers from October to May. **Spikelets** up to 7 mm long, occurring in pairs of which one spikelet is sessile and the other pedicellate, each with a delicate twisted awn. **Leaf blade** up to 8 mm wide, flat, relatively thick and densely hairy. **Leaf sheath** round and usually densely hairy. **Ligule** a brown membrane.

Habitat Common in open and mountainous grassland in reasonably high-rainfall areas. Grows on slopes, particularly in underutilized veld or in veld not regularly burnt. **Biomes**: Grassland and Savanna.

General An unpalatable grass which is eaten only early in spring and thereafter becomes hard an unacceptable to grazers, but it is still more acceptable than most other sourveld grasses. *Eulalia villosa* is sometimes confused with *Ischaemum fasciculatum* which occurs in moist areas and has a rhizome. Preferred by oribi. **Grazing value** average. **Ecological status**: Increaser I.

EUSTACHYS PASPALOIDES

Fan Grass
Bruin-hoenderspoor

A tufted perennial with short decumbent rhizomes and occasionally short stolons. Culms upright, up to 0,7 m tall, with purple nodes. **Inflorescence** consisting of 3—7 racemes, arranged digitately at the tip of the culm. Flowers from December to April. **Spikelets** up to 2 mm long, with slightly unequal glumes, arranged unilaterally in two rows on the rachis. A short, inconspicuous awn is present. **Leaf blade** up to 5 mm wide, glabrous, with a distinct midrib. **Leaf sheaths** flattened, overlapping each other, that giving a typical fan-like appearance to the leaf bases, glabrous, with shades of purple. **Ligule** an inconspicuous ring of hairs.

Habitat Generally occurs in open grassland and on stony, sandy soil, but also grows in bushveld and fynbos and sometimes on clay soil. **Biomes**: Grassland, Savanna and Fynbos.

General A palatable grazing grass which retains its palatability during winter and is well utilized by grazers. Although the grass can withstand heavy grazing, it nevertheless tends to disappear in natural veld during injudicious grazing practices. Apparently unpalatable in Mozambique. *Eustachys paspaloides* can be confused with *Chloris gayana*, but the latter is distinguished by its longer awns (up to 10 mm). **Grazing value** very high (low in eastern lowveld). **Ecological status**: Decreaser.

BOTHRIOCHLOA INSCULPTA

Pinhole Grass
Stippelgras

A tufted perennial with culms up to 1,0 m tall and a ring of white hairs around the nodes, often forming stolons. **Inflorescence** consists of 3—15 (rarely 20) spike-like racemes, up to 85 mm long, arranged semidigitately on the primary rachis. Flowers from October to June. **Spikelets** of two kinds, the one pedicellate and without an awn, the other sessile, with an awn up to 20 mm long and the glumes clearly pitted (hence the common name). **Leave blade** up to 8 mm wide, usually flattened. **Leaf sheath** round. **Ligule** a papery membrane with a hairy margin.

Habitat On all well drained, fertile soil types in open grassland or bushveld. Usually dominant when it colonizes disturbed sites. **Biomes**: Savanna and Grassland.

General An unpalatable, aromatic grass which tends to become hard and fibrous later in the season. Makes good hay, however. When this grass colonizes disturbed sites, it is difficult to replace it with better grazing by using normal grazing practices. It protects and stabilizes erodible soil. Sometimes confused with *Bothriochloa radicans*, but the glumes of the latter are not pitted. Also confused with *Dichanthium annulatum* var. *papillosum* and *Dichanthium aristatum*, but can be distinguished by the absence of an aroma and by the unpitted glumes in both those species. **Grazing value** low. **Ecological status**: Increaser IIb or IIc.

BOTHRIOCHLOA RADICANS

Stinking Grass
Stinkgras

A tufted perennial with branched culms up to 0,7 m tall, often with stolons and sometimes roots on the lower nodes. **Inflorescence** consists of 5—16 hairy racemes up to 70 mm long, arranged semidigitately on the primary axis. Flowers from October to April. **Spikelets** with glumes not pitted, in pairs of which one member is sessile, up to 5 mm long, with an awn up to 12 mm long and hairy, while the other is pedicellate, awnless and glabrous. **Leaf blade** up to 6 mm wide, usually glabrous and flattened. **Leaf sheath** glabrous. **Ligule** a ring of fine white hairs.

Habitat Common in warm, dry regions. Generally near vleis and other low-lying areas, or on stony slopes and around termite hills. Grows on most soil types but prefers heavy clay soils. **Biomes**: Savanna and Grassland.

General An unpalatable grass which is usually poorly utilized. Where it forms thick stands, it may be regarded as an indicator of disturbed veld. *Bothriochloa radicans* can be distinguished from *Bothriochloa insculpta* and *Bothriochloa bladhii* by the unpitted glumes and the shrubby growth form of *Bothriochloa radicans*. **Grazing value** very low to low. **Ecological status**: mostly Increaser IIb.

DICHANTHIUM ANNULATUM var. PAPILLOSUM

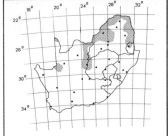

Vlei Finger Grass
Vleivingergras

A tufted perennial with a short rhizome, culms up to 1,0 m tall, with a ring of hairs around the nodes. **Inflorescence** consists of 3—6 (occasionally more) racemes up to 50 mm long, arranged semidigitately on the primary axis. Flowers throughout the year but mostly in late summer. **Spikelets** occur in pairs of which one member is sessile, up to 5 mm long, with an awn up to 25 mm long, and the other is pedicellate and awnless. **Leaf blade** up to 7 mm wide, flattened, glabrous, or sparsely hairy, with thickened leaf margins. **Leaf sheath** glabrous. **Ligule** a membrane, up to 4 mm long.

Habitat Occurs on heavy soils, often in moist areas such as vleis and riverbanks. **Biomes**: Savanna and Nama-Karoo.

General A reasonably palatable grass with a high leaf production. It is often confused with species of *Bothriochloa* but can be distinguished by the hairy spikelet pedicels of *Dichanthium annulatum* and the unpleasant smell and taste which the *Bothriochloa* species usually have.

* DICHANTHIUM ARISTATUM

Rainbow Vlei Grass
Reënboogvleigras

A tufted perennial with culms up to 1,1 m long, sometimes rooting from the lower nodes. **Inflorescence** consists of 2—5 racemes, up to 90 mm long, arranged semi-digitately on the primary axis, culm velvety below inflorescence. Flowers from October to June. **Spikelets** occur in pairs, one sessile, up to 4,5 mm long and with an awn up to 25 mm long, and the other pedicellate and awnless. **Leaf blade** up to 7 mm wide, flattened, glabrous to densely hairy. **Leaf sheath** shorter than internodes. **Ligule** a membrane with a hairy margin.

Habitat Usually occurs in disturbed areas such as roadsides and uncultivated lands. Often found in moist areas, on all soil types, with a preference for heavier soils. **Biome**: Savanna.

General Reasonably palatable with a medium leaf production. Distinguished from *Dichanthium annulatum* and species of *Bothriochloa* by the velvety appearance of the culm directly below the inflorescence. Preferred by blesbok.

ELEUSINE CORACANA subsp. AFRICANA

Goose Grass
Osgras

A tufted annual to weak perennial with culms up to 0,6 m tall. **Inflorescence** consists of 3—10 spikes up to 150 mm long, arranged digitately or subdigitately on the primary axis. The flowering culm remains green when the grass matures, compared with the straw colour of most other grasses. Flowers from October to May. **Spikelets** up to 8 mm long, sessile, awnless and strongly flattened, glumes almost equal. **Leaf blade** up to 10 mm wide, smooth, shiny and distinctly keeled. **Leaf sheath** compressed, with hairy margins. **Ligule** a membrane with sparse, scattered hairs.

Habitat Occurs in disturbed areas on all soil types. Found on compacted soils where few other species can survive. **Biomes**: Savanna, Grassland and Fynbos.

General This grass is the most common grass weed in cultivated lands in South Africa. Owing to its extensive and vigorous root system, it is difficult to control. It is utilized by livestock in the young stage, but becomes tough and unpalatable as it matures. The 'seeds' are used to make flour and cereal in time of food scarcity in some African countries. *Eleusine coracana* subsp. *africana* can be distinguished from *Dactyloctenium aegyptium* by the absence of an awn in the spikelets of the former. **Grazing value** low. **Ecological status**: Increaser IIc.

ISCHAEMUM AFRUM

Turf Grass
Turfgras

A tufted perennial with a strong rhizome and culms up to 1,5 m tall (rarely taller than 2 m). **Inflorescence** up to 200 mm long, consisting of 2—5 racemes arranged digitately to semidigitately on the primary axis, breaking up when mature. Flowers from October to April. **Spikelets** occur in pairs of which one spikelet is sessile and up to 8 mm long, the other pedicellate, both with an inconspicuous awn up to 10 mm long. **Leaf blade** up to 8 mm wide, glabrous, flat, with scabrid leaf margins. **Leaf sheath** round. **Ligule** a membrane up to 6 mm long.

Habitat Generally in heavy black turf, usually near water. Typical of hot low-lying areas with low to moderate rainfall. **Biomes**: Savanna and Grassland.

General Because of its strong rhizome it may create a problem when lands are made in turf. Used for plaiting hats. *Ischaemum afrum* could be confused with certain species of *Andropogon*, particularly *Andropogon schirensis*, but can be distinguished by its strong rhizome, the short, bent awn on each spikelet and its preference for turf. Preferred by reedbuck. **Grazing value** average. **Ecological status**: Decreaser.

ISCHAEMUM FASCICULATUM

Red Vlei Grass
Rooivleigras

A perennial with a strong rhizome and culms up to 1,3 m tall. **Inflorescence** up to 130 mm long, consisting of 2—5 racemes, arranged digitately or semidigitately on the primary axis. Flowers from November to May. **Spikelets** occur in pairs of which one member is 6 mm long and sessile, and the other long-pedicellate, each with an awn up to 10 mm long. **Leaf blade** up to 16 mm wide, flat, densely hairy to glabrous. **Leaf sheath** round. **Ligule** membranous.

Habitat Usually in wet areas such as vleis and riverbanks, often growing in water. Usually grows on heavy clay soils. **Biomes**: Grassland and Savanna.

General An average pasture grass. Sometimes confused with *Eulalia villosa*, but the latter rarely occurs in wet areas, does not have a strong rhizome and the awns are longer (up to 20 mm) than those of *Ischaemum fasciculatum*. Preferred by hippopotami.

UROCHLOA OLIGOTRICHA

Perennial Signal Grass
Meerjarige Beesgras

A tufted perennial with short rhizomes and culms up to 1,0 m tall. **Inflorescence** consists of up to 20 unilateral racemes up to 100 mm long, arranged on the primary axis. Flowers from December to May. **Spikelets** up to 5 mm long, usually in groups of 2—4, glabrous or hairy, lanceolate. **Leaf blade** up to 12 mm wide, flattened and densely to sparsely hairy. **Leaf sheath**: basal sheaths densely hairy, old sheaths break up into fibres. **Ligule** a ring of hairs.

Habitat Occurs in disturbed places such as uncultivated lands and along roads in grassland and bushveld, often in moist areas on clay and loam. **Biomes**: Savanna and Grassland.

General A palatable grass with a relatively high leaf production, well utilized by grazers. Can be considered as an indicator of disturbance. Distinguished from *Urochloa mosambicensis* and *Urochloa stolonifera* by the lanceolate spikelets and fibrous old leaf sheaths of *Urochloa oligotricha*.

ACROCERAS MACRUM

Nile Grass
Nylgras

A tufted perennial with a rhizome and culms up to 1,1 m tall. **Inflorescence** up to 200 mm long, consisting of 2—5 racemes, arranged singly on the primary axis. Flowers from November to July. **Spikelets** up to 5 mm long, glabrous, awnless, light green, with a blunt apex. **Leaf blade** up to 10 mm wide, flattened, glabrous or inconspicuously hairy, bright green and rounded at the base. **Leaf sheath** glabrous and papery when dry. **Ligule** an inconspicuous ring of short hairs.

Habitat Usually associated with vleis, riversides and other damp places where the grass grows in deep, fertile soils. **Biome**: Savanna.

General Nile Grass is a very palatable grass and well utilized by grazers. Successfully applied as an artificial pasture, particularly on damp soils. It makes good hay and reacts favourably to application of fertilizer. Propagated only by cuttings at present. The cultivar 'Cedara Select' is disease-resistant and has a high crude protein content. Once the grass is established, it is difficult to eradicate mechanically. **Grazing value** high to very high. **Ecological status**: Decreaser.

BEWSIA BIFLORA

False Love Grass
Vals-eragrostis

A tufted perennial, sometimes with a short rhizome, with erect unbranched culms up to 0,9 m tall (rarely 1,2 m). **Inflorescence** up to 200 mm long, with up to 12 racemes arranged irregularly on the primary axis. Flowers from November to April. **Spikelets** up to 9 mm long, shortly pedicellate, with 3 or 4 florets, lemmas with awns (up to 8 mm long) which develop 1—2 mm from the tip. **Leaf blade** up to 5 mm wide, flat, with a rough margin. **Leaf sheath** round, basal sheaths usually with purple shades. **Ligule** membranous, sometimes with hairs on both sides.

Habitat Common in open grassland or open mixed and sour bushveld. Prefers stony slopes and plains. Sometimes also found on riverbanks, along vleis and drainage canals. Grows on most soil types. **Biomes**: Grassland and Savanna.

General A tough grass with a low leaf yield. Rarely an important component of natural grassland. Of ornamental value in grass gardens. Sometimes confused with species of *Eragrostis*. Can be distinguished by the position of the awns, which develop 1—2 mm from the tips of the lemmas in *Bewsia biflora*. **Grazing value** low.

BRACHIARIA BRIZANTHA

Common Signal Grass
Broodsinjaalgras

A tufted perennial with culms erect or geniculate and up to 2,0 m long. **Inflorescence** consisting of 2—8 racemes up to 100 mm long, arranged on the primary axis and often hairy at the junction of raceme and axis. Flowers from December to April. **Spikelets** up to 6 mm long, hard, shiny and glabrous or sometimes hairy. **Leaf blade** up to 20 mm wide, flattened, glabrous or hairy, tapering to a sharp point and often with a thickened, purple leaf margin. **Leaf sheath** rounded and hairy. **Ligule** a ring of short dense, white hairs.

Habitat Grows on all soil types, with a preference for sandy soils in regions of moderate to high rainfall. Usually found in undisturbed veld, especially under trees in bushveld, but also on disturbed soils along roads and around termitaries. **Biomes**: Savanna, Grassland and Nama-Karoo.

General A reasonably palatable leafy pasture grass which apparently can withstand heavy grazing, but becomes woody late in the season. Common names such as 'Bread Grass' and 'Broodgras' remain a mystery, but could indicate that under certain circumstances the 'seeds' were used for grain. **Grazing value** average.

BRACHIARIA ERUCIFORMIS

Sweet Signal Grass
Litjiesinjaalgras

A tufted annual with erect or geniculate culms up to 0,6 m long, often rooting from the lower nodes. **Inflorescence** up to 70 mm long (occasionally up to 100 mm), consisting of a number of racemes, up to 30 mm long, arranged on the primary axis. Flowers from November to May. **Spikelets** up to 3 mm long, solitary, pedicellate and arranged in two rows on the rachis. **Leaf blade** up to 6 mm wide, often inconspicuously hairy on upper and lower sides and with a scabrous margin. **Leaf sheath** rounded, with short hairs. **Ligule** a ring of delicate white hairs.

Habitat Common but not exclusively on heavy clay or loam soils. Usually found in moist spots in disturbed areas such as gardens, irrigated lands and overgrazed vleis. **Biomes**: Savanna, Grassland and Nama-Karoo.

General A grass of average palatability with a low yield. Under natural conditions it is an indicator of waterlogged soil. A weed in gardens, irrigated lands and other disturbed areas. **Grazing value** low. **Ecological status**: Increaser IIc.

BRACHIARIA MARLOTHII

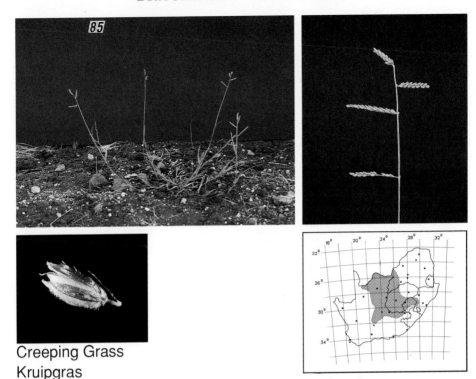

Creeping Grass
Kruipgras

A creeping or tufted annual to weak perennial with stolons, culms up to 0,5 m tall. **Inflorescence** up to 50 mm long, consisting of 1—6 (rarely more) racemes, up to 50 mm long, arranged irregularly on the primary axis. Flowers from December to May. **Spikelets** up to 2,5 mm long, unilateral and arranged in four rows. **Leaf blade** up to 5 mm wide, flattened, hairy, with thickened leaf margin. **Leaf sheath** hairy or glabrous. **Ligule** a ring of hairs.

Habitat   Common in overgrazed or disturbed areas such as gardens, lawns, roadsides and sometimes in cultivated lands, particularly where enough moisture is available. Usually grows in shallow sandy to loamy soils. **Biomes**: Savanna, Grassland and Nama-Karoo.

General   A palatable grass, but with a low leaf production. Regarded as particularly suitable pasture for sheep. Could become a troublesome weed in gardens and lawns. 'Seeds' sometimes cling to the wool of sheep.

BRACHIARIA NIGROPEDATA

Black-footed Signal Grass
Swartvoetjiegras

A tufted perennial with a creeping rhizome and unbranched culms up to 1,2 m tall. **Inflorescence** up to 100 mm long, consisting of 5—10 racemes up to 40 mm long, which are arranged singly on the primary axis. Flowers from November to April. **Spikelets** up to 4 mm long, arranged in two rows, hairy, with a typical black pedicel. **Leaf blade** up to 9 mm wide, flat, hairy, with a scabrid leaf margin. **Leaf sheath:** basal sheaths with velvety hairs. **Ligule** a ring of white hairs.

Habitat Generally on stony and rocky slopes in bushveld areas. Grows on most soil types, but prefers well drained sandy and stony soils. **Biomes**: Savanna and Nama-Karoo.

General A palatable grass with a high leaf production, which is well utilized by grazers. Regarded as an indicator of veld in good condition. One of the first grasses to disappear from the veld as a result overgrazing. Distinguished from other species by the typical short black pedicel of every spikelet (see common name). Preferred by sable antelope. **Grazing value** very high. **Ecological status**: Decreaser.

BRACHIARIA SERRATA

Velvet Signal Grass
Fluweelsinjaalgras

A tufted perennial with culms up to 0,9 m tall. **Inflorescence** consists of 2—10 unilateral velvety racemes up to 25 mm long, arranged on a hairy primary axis. Flowers from October to May. **Spikelets** up to 4,5 mm long, covered with dense white to purple hairs at especially the tips. **Leaf blade** up to 10 mm wide, hairy, with a thickened and sinuate leaf margin and a hard sharp tip. **Leaf sheath** round. **Ligule** a ring of short hairs.

Habitat Usually occurs in rocky places in open bushveld or grassland in a good condition. Also adapted to a wide range of other habitats. Prefers shallow sandy soils. **Biomes**: Grassland, Savanna and Fynbos.

General A relatively palatable grass with an average leaf production. Can be regarded as an indicator of veld in a good condition. *Brachiaria serrata* is distinguished from *Brachiaria arrecta* and *Brachiaria xantholeuca* by its sinuate leaf margin and the concentration of hairs at especially the tips of the spikelets. **Grazing value** average to high. **Ecological status**: Decreaser.

COELACHYRUM YEMENICUM

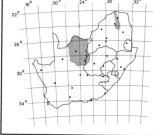

Small Scale Grass
Kortbeenskubgras

A tufted annual to weak perennial with erect or geniculate culms, up to 0,65 m long. **Inflorescence** up to 150 mm long, consisting of 2—8 contracted, spike-like racemes up to 70 mm long, arranged irregularly on the primary axis. Flowers from February to June. **Spikelets** up to 10 mm long, awnless, glumes glabrous, lemmas hairy at the base. **Leaf blade** up to 5,5 mm wide, glabrous, flattened or folded. **Leaf sheath** glabrous, except for sparse hairs at the mouth of the sheath. **Ligule** a membrane with a hairy margin.

Habitat Occurs on shallow limy soils or on calciferous pans in warm dry regions, generally in light shade. **Biomes**: Savanna and Nama-Karoo.

General Probably palatable, but with a variable leaf yield. Rarely abundant in natural veld. The inflorescence corresponds with that of *Diplachne eleusine*, which sometimes shares the same habitat. Can be distinguished by *Diplachne eleusine* being taller (up to 1,2 m) with culms that are more lax.

DIPLACHNE ELEUSINE

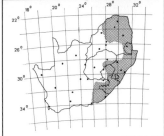

Large Scale Grass
Langbeenskubgras

A tufted perennial with a short rhizome and geniculate culms, up to 1,2 m tall. **Inflorescence** consists of 2—8 lax spikes, arranged irregularly on the primary axis. Flowers from November to April. **Spikelets** up to 8 mm long and overlapping. **Leaf blade** up to 4 mm wide and flattened. **Ligule** a membrane up to 1,5 mm long, with a hairy margin.

Habitat Generally in shade under trees, on sandy soils and sometimes on peat soils, often on stony slopes. **Biomes**: Savanna and Grassland.

General Probably has an average grazing value, but rarely abundant in natural veld. In other *Diplachne* species the ligule is a membrane without a hairy margin. *Diplachne eleusine* is sometimes confused with *Coelachyrum yemenicum* but can be distinguished by the longer and more of the former.

ECHINOCHLOA COLONA

Jungle Rice
Watergras

A tufted annual with culms up to 0,6 m tall (occasionally up to 1,0 m), often rooting from the lower nodes. **Inflorescence** up to 150 mm long, consisting of 6—15 spike-like, unilateral racemes up to 25 mm long of which the spikelets are arranged in four rows. Flowers from January to April. **Spikelets** up to 3 mm long, awnless, acute and inconspicuously pubescent. **Leaf blade** up to 8 mm wide, flat, glabrous. **Leaf sheath** glabrous. **Ligule** absent.

Habitat Associated with moist places such as pans, vleis, riverbanks and flood-plains. Often found in overgrazed or trampled sites. Grows on sandy loam to clayey soils. Sometimes occurs in gardens and other disturbed places. **Biomes**: Savanna, Grassland and Nama-Karoo.

General A palatable grass with a variable leaf production which depends on certain environmental conditions. A worldwide weed, occurring in more than sixty countries. Together with *Echinochloa crus-galli*, it can be a problem weed in South Africa, especially in rice fields. The grain is used as cereal. *Echinochloa colona* can cross-breed with *Echinochloa crus-galli*.

ERAGROSTIS ECHINOCHLOIDEA

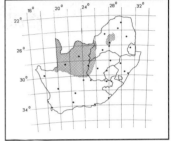

Tick Grass
Bosluisgras

A tufted perennial with culms up to 0,9 m tall, either upright or geniculate. **Inflorescence** a firm panicle, up to 160 mm long, with densely packed spikelets on solitary branches. Flowers from November to May. **Spikelets** up to 6 mm long, shortly pedicellate and strongly compressed. **Leaf blade** up to 6 mm wide, flattened or folded, hairy or glabrous. **Leaf sheath** keeled and papery. **Ligule** a ring of hairs.

Habitat Usually in disturbed areas such as roadsides and uncultivated lands, often around pans. Prefers shallow limy soils and sandy soils.
Biomes: Savanna and Nama-Karoo

General A grass of medium palatability, well utilized, particularly in the green stage. Can be regarded as an indicator of disturbed veld.
Eragrostis echinochloidea may be confused with *Eragrostis obtusa* and *Eragrostis brizantha*, but can be distinguished by the glumes of those two species which are usually obtuse, as opposed to the acute glumes of *Eragrostis echinochloidea*.
Grazing value low to average. **Ecological status**: Increaser IIb.

OPLISMENUS HIRTELLUS

Basket Grass
Bosgras

A perennial grass, with trailing culms up to 300 mm long (rarely 0,9 m) and rooting at the lower nodes. **Inflorescence** up to 100 mm long, consisting of 3—7 racemes, up to 30 mm long, arranged singly on the primary axis. Flowers from January to June. **Spikelets** up to 4 mm long (excluding awns), pedicellate, both glumes tapering to two sticky purple awns of which the longer is up to 15 mm long. **Leaf blade** up to 20 mm wide, lanceolate and often finely hairy. **Leaf sheath** with short white hairs on the margin. **Ligule** a short membrane with a hairy margin.

Habitat Generally in dense shade in mountain and coastal forests. It is often the only grass that grows in these habitats. **Biomes**: Savanna and Forest.

General Has a low leaf production. A certain variety is grown and sold as pot plants by florists in the U.S.A. Often confused with the closely related *Oplismenus undulatifolius* since they grow in the same habitat and intermediates sometimes occur. The spikelets of *Oplismenus undulatifolius* are clumped and not widely spaced as in *Oplismenus hirtellus*.

* PASPALUM DILATATUM

Common Paspalum
Gewone Paspalum

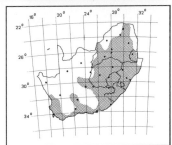

A tufted perennial often with a short rhizome, culms up to 1,5 m tall. **Inflorescence** up to 200 mm long, consists of 3—8 drooping racemes arranged on a primary axis with long white hairs at the junction of branches. Flowers from November to February. **Spikelets** up to 4 mm long, flat, with soft hairs. **Leaf blade** up to 15 mm wide, glabrous, except near base, with a rough margin. **Leaf sheath** glabrous, except basal sheaths. **Ligule** a conspicuous, papery membrane.

Habitat Common in damp places such as vleis, marshes, riverbanks and roadsides. Also in irrigated lands and orchards. Occurs on sandy loam to clayey soils. **Biomes**: Grassland, Savanna, Nama-Karoo and Fynbos.

General Originated in South America. A palatable to very palatable grass which is planted as pasture, particularly on damp soils. Used for grazing, hay and silage and in combination with clover. Reasonably resistant to heavy grazing. *Paspalum dilatatum* is sometimes confused with *Paspalum urvillei*, but the latter can be distinguished by its dense tuft and erect culms (up to 2,5 m tall) and the dense hairs, occasionally even resembling prickly-pear thorns, at the base of the plant. **Grazing value** high. **Ecological status**: Increaser IIb under natural conditions.

PASPALUM SCROBICULATUM

Veld Paspalum
Veldpaspalum

A tufted perennial with culms up to 0,9 m tall (usually much shorter) and often rooting at lower nodes. **Inflorescence** consists of 2—5 (rarely more) racemes up to 80 mm long, arranged singly on the primary axis. Flowers from October to May. **Spikelets** up to 3 mm long, rounded and glabrous. **Leaf blade** up to 8 mm wide, glabrous, often tinged with purple. **Leaf sheath** flattened and hairy, particularly the basal leaf sheaths. **Ligule** membranous.

Habitat Associated with wet soils near vleis or water furrows. In high rainfall regions the grass occurs on disturbed soils, for example along roadsides and in uncultivated lands. Has a preference for sandy soils and loam. **Biomes:** Grassland, Savanna and Fynbos.

General A palatable species with an average production. Planted as pasture in some parts of the world. A good pasture grass, but rarely an important component of natural veld. Sometimes confused with *Paspalum notatum* which has a strongly developed rhizome. **Grazing value** average. **Ecological status**: mostly Increaser IIc.

TETRACHNE DREGEI

Robies Cocksfoot
Kropaargras

A tufted perennial with a rhizome and culms up to 0,8 m tall. **Inflorescence** consists of a number of unilateral, relatively dense spikes, up to 40 mm long, arranged irregularly on the primary axis. Flowers from November to March. **Spikelets** up to 6 mm long, flattened, glabrous and awnless. **Leaf blade** up to 8 mm wide, glabrous, rolled and slightly curled. **Leaf sheath** hairy on the margin. **Ligula** a ring of dense hairs.

Habitat Generally on riverbanks, mountain slopes and rocky outcrops, at heights of more than 1 250 m above sea-level. Usually grows on sandy soils. **Biomes**: Nama-Karoo and Grassland.

General A very palatable species, probably the most valuable in its distribution range. Presently planted on a small scale as pasture. A semi-prostrate grass which can form dense stands if it is optimally grazed. **Grazing value** very high. **Ecological status**: Decreaser.

UROCHLOA MOSAMBICENSIS

Bushveld Signal Grass
Bosveldbeesgras

A loosely tufted perennial with stolons and culms up to 1,5 m tall and sometimes roots at the lower nodes. **Inflorescence** consists of 3—12 racemes, up to 100 mm long, arranged along the primary axis. Flowers from October to May. **Spikelets** up to 5 mm long, with short pedicels and arranged unilaterally. **Leaf blade** up to 20 mm wide, flattened, generally hairy and relatively broad close to ligule. **Leaf sheath** usually hairy, with a ring of hairs around the nodes. **Ligule** a ring of hairs.

Habitat Usually occurs in disturbed areas such as overgrazed and trampled veld and particularly along roadsides. Grows on most fertile soil types. **Biomes**: Savanna and Grassland.

General A palatable pasture grass of medium leaf production and well grazed by stock. Can sometimes be seen as an indicator of disturbed veld, often as a result of long droughts or overgrazing. Continuous selective grazing of this highly palatable grass can cause a serious decline in veld conditions, and careful veld management is therefore needed where the grass is dominant. Confused with the closely related *Urochloa stolonifera* which is a smaller plant (culms up to 300 mm tall) with spikelets which are untidily arranged. Forms intermediates with *Urochloa oligotricha*, but can be distinguished from the latter by its lanceolate spikelets and fibrous old leaf sheaths. Preferred by white rhinoceros, hippopotami and impala. **Grazing value** mostly high. **Ecological status**: Increaser IIc.

UROCHLOA PANICOIDES

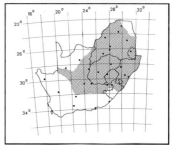

Garden Signal Grass
Tuinbeesgras

A tufted annual with culms sometimes branched and up to 0,75 m tall, rooting at the lower nodes. **Inflorescence** of 2—7 racemes, up to 60 mm long and arranged singly on the primary axis. Flowers from October to May. **Spikelets** up to 5 mm long, with short pedicels, mostly glabrous and arranged unilaterally. **Leaf blade** up to 18 mm wide, relatively short and broad, hairy and with a wavy, thickened margin. **Leaf sheath** hairy. **Ligule** a hairy ring.

Habitat Common in disturbed areas such as cultivated fields, gardens and roadsides. Often found in moist places where water accumulates. Also occurs in patches on compacted bare patches. Grows on most soil types, with a preference for heavy and fertile soils. **Biomes**: Savanna, Grassland and Nama-Karoo.

General A palatable grass with a low leaf production under natural conditions. Uncommon in natural veld, but a common weed which can form dense stands in and around cultivated fields, gardens and other disturbed areas. Often plays an important role during the stabilization of bare compacted soils. Can be distinguished from other annual *Urochloa* species by its short lower glumes (not more than half the length of the spikelet) and its glabrous spikelets. **Grazing value** low. **Ecological status**: Increaser IIc.

PANICULATE INFLORESCENCES

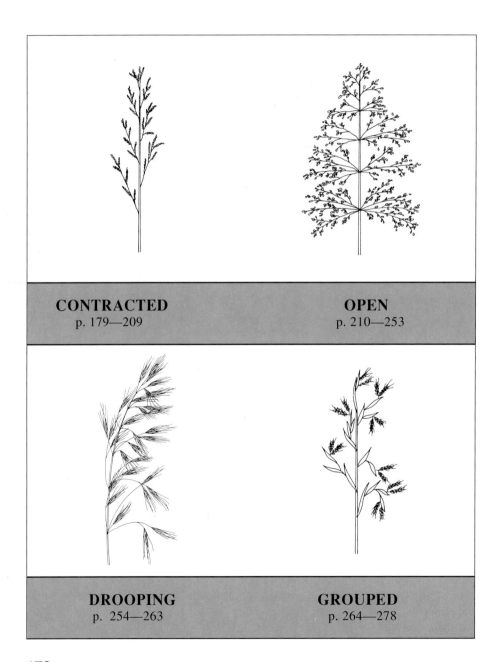

CONTRACTED
p. 179—209

OPEN
p. 210—253

DROOPING
p. 254—263

GROUPED
p. 264—278

ARISTIDA CANESCENS subsp. CANESCENS

Pale Three-awn
Vaalsteekgras

A tufted perennial, with culms usually unbranched and up to 1,0 m tall (rarely up to 1,5 m). **Inflorescence** a slightly contracted panicle, up to 200 mm long, generally with the spikelets borne at the tip of branches. Flowers from December to May. **Spikelets** up to 22 mm long (awns included), with three awns, the central one slightly longer than the other two. **Leaf blade** up to 2 mm wide, curled when old, hard, with scabrid leaf margins.

Habitat Usually in bushveld and open grassland, in disturbed and eroded areas. Grows in poor, shallow, sandy and stony soils. Often found on bare compacted soils. **Biomes**: Grassland and Savanna.

General Owing to its low productivity and the hardness of its leaves, this grass has hardly any grazing value. Certain forms of *Aristida canescens* subsp. *canescens* and *Aristida junciformis* subsp. *junciformis* may be confused, but *Aristida junciformis* subsp. *junciformis* can be distinguished by its lower glume which is either tapered or awned and the plant is also more densely tufted. **Grazing value** very low. **Ecological status**: Increaser IIc.

ARISTIDA RHINIOCHLOA

Large-seeded Three-awn
Skurwesteekgras

A tufted annual with erect, branched culms up to 1,0 m tall. **Inflorescence** an open to slightly contracted panicle up to 300 mm long. Flowers from January to May. **Spikelets** up to 50 mm long (including awns), with three awns, scabrous, without a column and articulation. **Leaf blade** up to 4 mm wide, keeled and scabrous, particularly the margin. **Leaf sheath** with long white hairs at the apex. **Ligule** a ring of short hairs.

Habitat Occurs in regions with a moderate to low annual rainfall. Prefers disturbed areas such as overgrazed veld, uncultivated lands and eroded soils. Often found on stony slopes. Grows in any soil type. **Biome**: Savanna.

General A very low grazing value as a result of a low yield and the hardness and prickliness of the leaves. Increases during long droughts and an indicator of overgrazing, drought or other disturbances. However, it plays a major role in the protection of erodible soils in low-rainfall areas. **Grazing value** very low. **Ecological status**: Increaser IIc.

CHRYSOPOGON SERRULATUS

Golden Beard Grass
Gouebaardgras

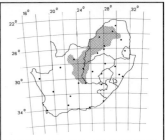

A tufted perennial, often with a short rhizome, with culms up to 1,0 m tall, generally unbranched. **Inflorescence** an open or contracted panicle, up to 200 mm long, consisting of whorls of lax branches on the primary axis. Flowers from December to April. **Spikelets** up to 8 mm long, in groups of three of which one is sessile and two are pedicellate and hairy, each with an awn up to 25 mm long. **Leaf blade** up to 10 mm wide, flat or folded and curling when old. **Leaf sheath** compressed and keeled. **Ligule** a membrane with a hairy margin.

Habitat Commonly on steep, dry, stony slopes or stony koppies. Often on dolomite. Prefers shallow stony soil, but also grows on clay and loam soil. **Biomes:** Savanna, Grassland and Nama-Karoo.

General A palatable to very palatable grass, with a high leaf production, well utilized by livestock and game. Used as thatching grass in certain parts of the western Transvaal. **Grazing value** very high. **Ecological status:** Decreaser.

DIGITARIA DIAGONALIS var. DIAGONALIS

Brown-seed Finger Grass
Bruinsaadvingergras

A tufted perennial with a bulbous base and erect unbranched culms up to 1,5 m tall. **Inflorescence** up to 300 mm long, with racemes up to 200 mm long, arranged singly or in groups on the primary axis. Flowers from December to April. **Spikelets** up to 2,5 mm long, in groups, dark brown to black, interspersed with white bristles. **Leaf blade** up to 3 mm wide, flat, often rough at the margin and generally glabrous. **Leaf sheath:** basal sheaths break up into fibres when old, usually finely hairy. **Ligule** membranous.

Habitat Common in open, sour grassland, on slopes and in moist areas such as vleis. Grows on most soil types, but prefers sandy soils. **Biomes:** Grassland and Savanna.

General An average pasture grass with a variable leaf production, which is reasonably well utilized by grazers. *Digitaria diagonalis* var. *diagonalis* could be confused with *Digitaria eylesii* which does not have white bristles between the spikelets, or with *Digitaria maitlandii* which has much shorter culms (up to 400 mm). **Grazing value** average.

DINEBRA RETROFLEXA var. CONDENSATA

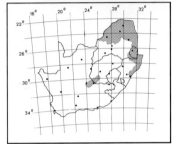

Catstail Vlei Grass
Katstertturfgras

A tufted annual, with culms up to 0,8 m tall. **Inflorescence** up to 300 mm long, consisting of a number of spikes up to 50 mm long, arranged on the primary axis. Flowers from December to May. **Spikelets** up to 9 mm long and glabrous, with glumes almost equal and up to 8 mm long. **Leaf blade** up to 8 mm wide, flattened, glabrous or sparsely hairy. **Ligule** a membrane.

Habitat Generally in disturbed, moist and wet areas, usually on black turf and waterlogged soils and sometimes in shade. **Biome**: Savanna.

General Probably of little grazing value owing to its low leaf production. Occurs as a weed in lands, particularly in rice-fields.

DIPLACHNE FUSCA

Swamp Grass
Kuilgras

A water-loving, tufted perennial with a rhizome, culms often branched, up to 1,5 m tall, sometimes rooting from the lower nodes. **Inflorescence** up to 400 mm long, consisting of numerous lax racemes up to 150 mm long, arranged irregularly and/or in groups on the primary axis. Flowers from October to April. **Spikelets** up to 14 mm long, not overlapping, with 5—10 florets. **Leaf blade** up to 5 mm wide, flat, smooth, glabrous, with thread-shaped tips. **Leaf sheath:** upper sheaths round and lower ones compressed. **Ligule** a conspicuous white membrane up to 5 mm long.

Habitat Grows in or near fresh or brackish water sources such as vleis, rivers, lakes and floodplains or pans. **Biomes:** Grassland, Savanna, Fynbos Nama-Karoo and Succulent Karoo.

General A valuable pasture grass in vleis and brackish soils, which also makes good hay. This grass is very variable in growth form. It can be distinguished from species of *Eragrostis* by its conspicuous white membranous ligule, as opposed to the hairy ligules of the *Eragrostis* species.

EHRHARTA VILLOSA var. MAXIMA

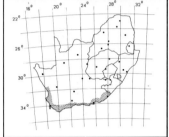

Dune Ehrharta
Doppiesgras

A tufted perennial, sometimes with a rhizome, with culms up to 1,5 m tall and often rooting at the lower nodes. **Inflorescence** a small, often unilateral panicle, up to 200 mm long. Flowers from September to March. **Spikelets** up to 18 mm long, glumes almost equal, lemmas long-haired and each ending in a short awn. **Leaf blade** up to 8 mm wide, flattened or rolled and often deciduous.

Habitat Usually grows on sand, often on coastal sand dunes and sometimes found up to 1 km from the coastline.

General Probably of little value for grazing, because the grass has a relatively low leaf production. However, it is an important sand binder and stabilizer. Two varieties are distinguished, namely *Ehrharta villosa* var. *maxima* and *Ehrharta villosa* var. *villosa*. The former is a more robust plant and has longer spikelets.

ENNEAPOGON CENCHROIDES

Nine-awned Grass
Negenaaldgras

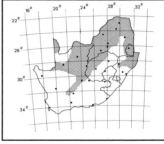

A tufted annual or weak perennial, with culms generally geniculate and up to 1,0 m tall. **Inflorescence** a tapering, contracted or open panicle, up to 200 mm long. Usually flowers from December to May. **Spikelets** up to 5 mm long, with lemmas typically bearing nine hairy awns. **Leaf blade** up to 8 mm wide, flattened, or rolled when dry, with dense short hairs and a rough margin. **Leaf sheath** rounded. **Ligule** a ring of short hairs.

Habitat Common in disturbed areas such as roadsides and overgrazed veld. Also found in natural veld after drought. Grows mostly in sandy soils and sometimes in limestone communities or heavier soils. **Biomes**: Savanna, Grassland and Nama-Karoo.

General A hardy species, able to withstand long droughts and heavy grazing. Of particular value in low-rainfall areas, where the grass can form thick stands, giving a relatively high yield. Can be regarded as an indicator of overgrazing and other disturbances. *Enneapogon cenchroides* is sometimes confused with *Cenchrus ciliaris* or *Schmidtia kalihariensis* which may occur in the same habitat. However, the inflorescence of *Cenchrus ciliaris* is a dense spike, as opposed to the panicle of *Enneapogon cenchroides*. *Schmidtia kalihariensis* has only five awns per spikelet and an unpleasant sour smell. **Grazing value** variable but usually low. **Ecological status**: Increaser IIc.

ENNEAPOGON SCOPARIUS

Bottlebrush Grass
Kalkgras

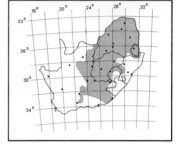

A tufted perennial with culms up to 0,6 m tall. **Inflorescence** a dense spike-like panicle up to 120 mm long. Flowers from November to May. **Spikelets** up to 4,5 mm long, with lemmas bearing 9 hairy awns. **Leaf blade** up to 3 mm wide, filiform and covered with fine hairs. **Leaf sheath** hairy. **Ligule** an inconspicuous ring of hairs.

Habitat Commonly occurs under trees in disturbed veld. Generally found on stony slopes and in limestone communities. Prefers shallow soils. **Biomes**: Savanna, Grassland and Nama-Karoo.

General A moderately palatable species which is poorly utilized, probably owing to its tough wiry leaves. Rarely dominant in natural veld. Confused with *Enneapogon cenchroides* which has a more open growth form, longer culms (up to 0,9 m tall), longer inflorescences (up to 150 mm), and leaves up to 8 mm wide. **Grazing value** low. **Ecological status**: mostly Increaser IIb.

ERAGROSTIS CAPENSIS

Heart-seed Love Grass
Hartjiesgras

A tufted perennial with culms up to 0,9 m tall. **Inflorescence** a sparsely branched panicle, up to 150 mm long, consisting of relatively few (15—30) large spikelets. Flowers from September to April. **Spikelets** up to 15 mm long, flattened and pedicels appressed to the primary axis, sometimes on short branches. **Leaf blade** up to 5 mm wide, glabrous, rolled, often tinged with purple at the base. **Leaf sheath** smooth and round. **Ligule** an inconspicuous ring of short hairs.

Habitat Common in open sour grassland on shallow soils. Elsewhere it occurs in moist areas, for example along vleis and on riverbanks and also in disturbed areas such as roadsides where it can form dense stands. Grows on any soil type, from sand to clay. **Biomes**: Grassland, Savanna and Fynbos.

General A reasonably palatable grass, but it has a low leaf production. One of the first grasses to sprout in the early spring and after a veld fire, and then well utilized. Often confused with *Eragrostis superba* or *Eragrostis racemosa*, but the spikelets of *Eragrostis superba* are typically toothed at the sides and not smooth like those of *Eragrostis capensis*, while the spikelets of *Eragrostis racemosa* are oblong and sharp-pointed, as opposed to the oval spikelets of *Eragrostis capensis*. Sheep often graze on the young inflorescences. **Grazing value** average. **Ecological status**: Variable.

ERAGROSTIS GUMMIFLUA

Gum Grass
Gomgras

A tufted perennial with culms up to 0,9 m tall, with a sticky substance at the nodes. **Inflorescence** a hard, slightly contracted panicle, up to 250 mm long. Flowers from November to April. **Spikelets** up to 4 mm long and glabrous, with glumes almost equal. **Leaf blade** up to 4,5 mm wide, young leaves initially flattened but later rolled. **Leaf sheath** shiny and sticky at the top, soil particles often clinging to the sticky substance. **Ligule** a membrane with a hairy margin.

Habitat Often more dominant in the high-rainfall areas of its distribution range. Prefers open grassland and open patches in bushveld. Sometimes found along roadsides. Prefers sandy soil and poor, leached stony soil. **Biomes**: Savanna, Grassland and Fynbos.

General
A hard and unpalatable species which is eaten only when hardly any other grazing is available. Gives rise to severe selective grazing problems within its distribution range. In Lesotho it is used for broom-making. *Eragrostis viscosa* is another sticky *Eragrostis* species, but it is annual and has hairy spikelets. Preferred by sable. **Grazing value** very low. **Ecological status**: mostly Increaser IIc.

ERAGROSTIS NINDENSIS

Wether Love Grass
Hamelgras

A tufted perennial with upright, unbranched culms up to 0,9 m tall. **Inflorescence** an open or slightly contracted panicle with a rather rigid primary axis, 50—280 mm long. Flowers from January to May. **Spikelets** up to 20 mm long, more or less equally broad throughout, with a sharp tip, hard and glabrous. **Leaf blade** up to 3 mm wide, usually with long soft hairs, colour varying from bright green to reddish brown later in the season. **Leaf sheath** hard and striped, hairy or glabrous. **Ligule** a ring of short, spaced hairs.

Habitat Occurs on many soil types, with a preference for stony, sandy soils, usually on bare patches. **Biomes**: Savanna and Nama-Karoo.

General A relatively palatable grass which tends to sprout early in the rainy season. Various common names refer to the connection between this grass and sheep, indicating that the grass is well utilized by them, particularly in semi-arid regions. Often confused with *Eragrostis racemosa* which has dark and shorter (up to 10 mm long) spikelets. **Grazing value** average. **Ecological status**: Increaser IIc.

ERAGROSTIS PALLENS

Broom Love Grass
Besemgras

A tufted perennial with culms up to 1,5 m tall (occasionally up to 2 m). **Inflorescence** an open or contracted panicle up to 250 mm long. Flowers from December to May. **Spikelets** up to 25 mm long, glabrous, shortly pedicellate, green to yellow. **Leaf blade** up to 8 mm wide, hard, usually glabrous, inrolled. **Leaf sheath** yellow when it turns older; basal sheaths often with silky hairs. **Ligule** a ring of hairs.

Habitat Generally in the sandveld parts of the bushveld. Grows in sandy soils and often in moist sandy soils near seasonal pans, often in association with *Terminalia sericea* and *Burkea africana* veld. **Biome**: Savanna.

General A hard unpalatable grass, poorly utilized by livestock. The inflorescences are sometimes utilized. Sometimes used for thatching. **Grazing value** low. **Ecological status**: mostly Increaser IIb.

ERAGROSTIS PLANA

Tough Love Grass
Taaipol-eragrostis

A tufted perennial with culms up to 1,0 m tall. **Inflorescence** a panicle up to 250 mm long, contracted at the tip and open at the base. Flowers from September to March. **Spikelets** up to 10 mm long, glabrous, with short, unequal glumes. **Leaf blade** up to 4 mm wide, smooth, usually glabrous and folded, leaves strong and difficult to break. **Leaf sheath** strong, flattened and overlapping at the base of the plant, giving it a typical fan-like appearance. **Ligule** a ring of short hairs.

Habitat Occurs on disturbed soil such as uncultivated lands, roadsides and trampled areas, such as those adjoining cattle-kraals and watering points. Prefers damp places, particularly in the drier parts of the country. **Biomes**: Savanna and Grassland.

General An unpalatable grass with tough leaves, generally poorly utilized by stock and game. The grass is utilized particularly in the young stage in the drier parts of the country. Leaves are used for weaving. *Eragrostis plana* could be confused with particularly *Sporobolus africanus* and *Sporobolus pyramidalis*, but can easily be distinguished by the small spikelets (up to 3 mm long) of the latter two species. *Eragrostis plana* is the only *Eragrostis* species with fan-shaped leaf bases. **Grazing value** low. **Ecological status**: mostly Increaser IIc.

ERAGROSTIS RACEMOSA

Narrow Heart Love Grass
Smalhartjiesgras

A tufted perennial with upright culms, up to 0,5 m tall. **Inflorescence** a panicle of up to 150 mm long, with 8—30 spikelets relatively shortly pedicellate. Flowers from August to May. **Spikelets** up to 10 mm long, slightly rough and usually olive-green. **Leaf blade** up to 5 mm wide, generally hairy, curls up and becomes reddish brown as it matures. **Leaf sheath** round and veined. **Ligule** an inconspicuous ring of short hairs.

Habitat Common in open grassland with a high rainfall, but also in bushveld, vleis and on riverbanks. Occurs on most soils, preferring shallow sandy soil. **Biomes**: Savanna, Grassland and Fynbos.

General A relatively palatable grass but with a low leaf production and therefore not an important grass for grazing. *Eragrostis racemosa* is sometimes confused with *Eragrostis nindensis* or *Eragrostis capensis*, but can be distinguished by *Eragrostis capensis* having considerably broader spikelets, and those of *Eragrostis nindensis* being longer and narrower than those of *Eragrostis racemosa*. **Grazing value** low. **Ecological status**: Increaser IIb or IIc.

ERAGROSTIS SUPERBA

Sawtooth Love Grass
Weeluisgras

A tufted perennial with geniculate or erect culms up to 1,0 m tall. **Inflorescence** an open or contracted panicle, up to 300 mm long. Flowers from August to May. **Spikelets** up to 16 mm long, glabrous, heart-shaped, margin flattened, conspicuously toothed. **Leaf blade** up to 12 mm wide, glabrous, usually rolled. **Leaf sheath** glabrous, keeled. **Ligule** an inconspicuous ring of white hairs.

Habitat Usually found in disturbed localities. Prefers sandy and stony soils but sometimes grows on heavier soils. **Biomes**: Grassland, Savanna and Fynbos.

General A reasonably valuable and palatable grass, with a variable leaf production. It is drought-resistant and in the U.S.A. it is planted as pasture and used for hay. Rarely an important component of the natural veld. Plays an important role in the stabilization of bare and erodible soils. Often confused with *Eragrostis capensis* but can be distinguished by the conspicuously toothed margin of its spikelets, while those of *Eragrostis capensis* are smooth. **Grazing value** average to low. **Ecological status**: Variable.

ERIOCHLOA STAPFIANA

Harpoon Grass
Harpoengras

A tufted perennial with culms up to 1,7 m tall. **Inflorescence** a contracted panicle, up to 250 mm long, usually with short hairs on the axis. Flowers from October to May. **Spikelets** up to 4 mm long, with dark pedicels, upper glume acute or bearing an awn up to 1 mm long, lemma with a sharp tip. **Leaf blade** up to 8 mm wide and usually flattened. **Leaf sheath** shortly hairy.

Habitat Generally grows on heavy soils in wet areas such as riverbanks. **Biome**: Savanna.

General A relatively unknown grass, but probably a reasonably good pasture. Closely related to *Eriochloa meyeriana* and intermediates can also occur. Can be distinguished by the spikelets of *Eriochloa meyeriana* which are less acute, and its culms which are often geniculate and rooting from the lower nodes.

LOUDETIA SIMPLEX

Common Russet Grass
Stingelgras

A tufted perennial with unbranched culms up to 1,2 m tall, the nodes of which generally have a ring of hairs. **Inflorescence** a contracted or open panicle up to 300 mm long. Flowers from November to January. **Spikelets** have a glabrous awn up to 50 mm long, glumes unequal, glabrous or hairy, with blunt tips. **Leaf blade** up to 5 mm wide, flattened or rolled, usually glabrous. **Leaf sheath** splits into fibres as it matures. **Ligule** an inconspicuous ring of short hairs.

Habitat Common in stable, open grassland on poor, shallow, sandy soils in the higher rainfall regions of the country. Also on a wide variety of other habitats such as stony slopes, edges of vleis and certain bushveld areas. **Biomes**: Grassland and Savanna.

General An unpalatable grass, poorly utilized by grazers. It is rarely an important component of natural veld and when it is common, it could indicate poor veld management. The inflorescences are sometimes bound together to make brooms. *Loudetia simplex* is sometimes confused with *Loudetia flavida*, a closely related species, occasionally occurring in the same habitat, but the glumes of *Loudetia flavida* are always sharply pointed, as opposed to the obtuse glumes of *Loudetia simplex*. **Grazing value** very low. **Ecological status**: variable Increaser.

MELINIS NERVIGLUMIS

Bristle-leaved Red Top
Steekblaarblinkgras

A dense tufted perennial with unbranched culms up to 1,2 m tall. **Inflorescence** a relatively dense, contracted or open panicle, up to 100 mm long. Flowers from November to September. **Spikelets** up to 5 mm long, dense and covered with white, pink or purple hairs up to 4 mm long. **Leaf blade** up to 3,5 mm wide, usually glabrous and rolled. **Leaf sheath:** basal sheaths strongly overlapping and hairy at the base, other sheaths hairy or glabrous. **Ligule** an inconspicuous hairy rim.

Habitat Generally on shallow stony soils in undisturbed, open grassland or bushveld. Usually on stony slopes. **Biomes**: Grassland, Savanna and Fynbos.

General A moderately palatable species of which the leaves can become hard and stringy, especially late in the season. *Melinis nerviglumis* is sometimes confused with *Melinis repens* subsp. *repens*. The former, however, is perennial, has a denser tuft with rolled leaf blades, and a denser contracted panicle. **Grazing value** average. **Ecological status**: variable but mostly Increaser I.

197

PENTASCHISTIS CURVIFOLIA

Tassel Grass
Kwasgras

A tufted perennial with culms up to 0,5 m tall. **Inflorescence** a dense contracted panicle, up to 80 mm long. Flowers from September to December. **Spikelets** up to 12 mm long, glumes ivory-coloured and lemmas bearing dark awns. **Leaf blade** up to 4 mm wide, leaf margin thickened, glabrous, initially flattened, but later folded and rolled from the margin, old leaves curled.

Habitat Found in a wide variety of habitat types at different altitudes, but usually on sandy soil. **Biome**: Fynbos.

General Probably of average grazing value. Smaller plants of *Pentaschistis curvifolia* could be confused with *Pentaschistis pallida*, but the latter has smaller spikelets (up to 5 mm long) and glands on most parts of the plant.

POGONARTHRIA SQUARROSA

Herringbone Grass
Sekelgras

A tufted perennial with culms up to 0,8 m tall (rarely 1,4 m). **Inflorescence** consists of a number of sickle-shaped racemes, arranged on a primary axis. Flowers from October to May. **Spikelets** up to 8 mm long with sharp-pointed glumes and lemmas. **Leaf blade** up to 5 mm wide, often inrolled and glabrous. **Leaf sheath** glabrous, round or keeled. **Ligule** a ring of short white hairs.

Habitat Generally in disturbed areas such as uncultivated lands and roadsides. Grows on most soils, but prefers poor sandy soils. Also scattered in undisturbed grassland. **Biomes**: Grassland, Savanna and Nama-Karoo.

General A hard unpalatable grass with a low yield, poorly utilized by grazers. An indicator of old lands and poor soils. Apparently toxic in certain parts of Mozambique. Inflorescence of *Pogonarthria squarrosa* could be confused with that of *Sporobolus pyramidalis*, but the spikelets differ considerably. **Grazing value** very low. **Ecological status**: mostly Increaser IIc.

SCHMIDTIA KALIHARIENSIS

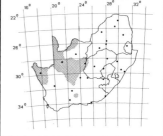

Sour Grass
Suurgras

A tufted annual, with culms up to 0,9 m tall, generally unbranched, often rooting from the lower nodes. **Inflorescence** an interrupted contracted to spike-like panicle, up to 150 mm long. Flowers throughout the year, but usually from mid- to late summer. **Spikelets** up to 17 mm long, hairy, lemma with 5 awns. **Leaf blade** up to 15 mm wide, densely hairy and abruptly tapering to a sharp tip. **Leaf sheath** densely hairy and round. **Ligule** an inconspicuous ring of hairs.

Habitat Generally in dry to very dry areas. Prefers sandy and disturbed locations. **Biomes**: Savanna, Succulent Karoo and Nama-Karoo.

General Relatively palatable before the flowering stage and as dry grass. The inflorescences have a high nutritive value and are well utilized. Also makes a good hay. During the flowering season the entire plant has an unpleasant sour odour and is slightly sticky. Can be regarded as an indicator of overgrazing and/or drought. May be confused with *Schmidtia pappophoroides*, but the latter is perennial, has smaller leaves (up to 7 mm wide), is less hairy and the characteristic sour smell and stickiness are lacking. **Grazing value** low. **Ecological status**: Increaser IIc.

SCHMIDTIA PAPPOPHOROIDES

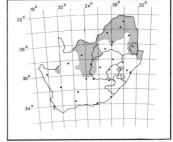

Sand Quick
Sandkweek

A tufted perennial, often with stolons and roots at the lower nodes, with culms up to 0,9 m tall. **Inflorescence** a spike-like contracted panicle, up to 120 mm long, with short pedicels. Usually flowers from October to June. **Spikelets** up to 15 mm long, with green glumes, and lemmas with 5 relatively straight awns up to 10 mm long. **Leaf blade** up to 7 mm wide, often covered with dense short hairs, usually involuted. **Leaf sheath** mostly round and hairy. **Ligule** a ring of hairs.

Habitat Generally in warm areas with relatively low rainfall. Prefers sandy and stony soils, but also found on limy and sometimes on stony clay soils. **Biomes**: Savanna, Nama-Karoo and Succulent Karoo.

General A palatable to very palatable grass with a limited leaf production. The grass is very variable in size and particularly in hairiness. Preferred by roan. Often confused with *Enneapogon cenchroides* which typically has nine lemma awns. **Grazing value** mostly high. **Ecological status**: mostly Decreaser.

SETARIA HOMONYMA

Fan-leaved Bristle Grass
Waaierblaarmannagras

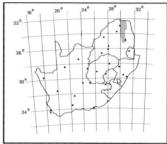

A tufted annual with culms up to 1,0 m tall, often branched, geniculate and rooting at the lower nodes. **Inflorescence** an open panicle, up to 200 mm long, consisting of 6—10 racemes, up to 60 mm long, spreading from or suberect on the primary axis. Flowers from December to May. **Spikelets** up to 3 mm long, glabrous, each with a stiff bristle attached to the pedicel. **Leaf blade** up to 35 mm wide, lanceolate, pleated lengthwise, scabrous along the leaf margin and often finely hairy. **Leaf sheath** compressed, with white hairs along the margin. **Ligule** a ring of hairs.

Habitat Occurs in bushveld and forests in middle and lowveld regions. Found mostly in the shade on moist soils in disturbed areas such as roadsides, riverbanks and flood plains. Sometimes a weed in cultivated lands. **Biome**: Savanna.

General Probably with an average grazing value. Occurs as a weed in disturbed areas. Distinguished from other annual species of *Setaria* with similar inflorescences by way of its pleated and lanceolate leaves.

SPOROBOLUS PYRAMIDALIS

Catstail Dropseed
Katstert-fynsaadgras

A tufted perennial with unbranched culms, up to 1,2 m tall, seldom taller. **Inflorescence** a contracted panicle up to 0,5 m long, consisting of a variety of straight branches, arranged irregularly on the primary axis. Flowers from November to May. **Spikelets** up to 2 mm long, glabrous and pointed. **Leaf blade** up to 10 mm wide, glabrous. **Leaf sheath** rounded and smooth. **Ligule** an inconspicuous ring of hairs.

Habitat Generally in relatively fertile disturbed areas such as trampled veld and uncultivated lands with a relatively high fertility. Prefers low-lying warm regions with high rainfall. Grows on all soil types. **Biomes**: Savanna and Grassland.

General A tough, unpalatable species and an indicator of overgrazed, trampled and disturbed areas. A troublesome weed in perennial planted pastures. It nevertheless plays an important role in controlling erosion in disturbed areas. *Sporobolus pyramidalis* could occasionally be confused with the closely related *Sporobolus africanus*, but the latter can be distinguished by its more contracted spike-like panicle. *Sporobolus pyramidalis* could also be confused with *Sporobolus fimbriatus*, but the inflorescence of the latter is more oval and its tip is not as sharply pointed. Also see *Eragrostis plana*. **Grazing value** low. **Ecological status**: mostly Increaser IIc.

STIPAGROSTIS CILIATA var. CAPENSIS

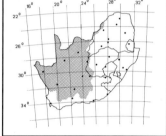

Tall Bushman Grass
Langbeenboesmangras

A tufted perennial with erect or geniculate culms, up to 1,0 m tall, nodes with a distinct ring of white hairs. **Inflorescence** a sparse open or contracted panicle, up to 300 mm long. Flowers from August to October and from February to June. **Spikelets** up to 12 mm long (awns excluded), staw-coloured, usually with a purple base and three awns of which the central one is plumose. **Leaf blade** up to 1,5 mm wide, usually curled, hard, leaves mainly basal. **Leaf sheath** hairy. **Ligule** a ring of short hairs.

Habitat Generally in sandveld, gravel plains and riverbeds. Usually grows on coarse sandy soil. **Biomes**: Savanna, Nama-Karoo and Succulent Karoo.

General A palatable species and a valuable pasture. The grass is resistant to drought, has a high nutritional value, even in the dry stage. It is also important in the binding of sand. Often occurs in association with *Stipagrostis obtusa* which has glabrous, dark nodes and is a smaller plant (up to 0,6 m high) than *Stipagrostis ciliata* var. *capensis*. Preferred by blue wildebeest. **Grazing value** high. **Ecological status**: Decreaser.

STIPAGROSTIS OBTUSA

Small Bushman Grass
Kortbeenboesmangras

A compact, tufted perennial with culms up to 0,6 m tall, usually with dark nodes. **Inflorescence** a contracted, interrupted panicle up to 200 mm long. Flowers from July to May. **Spikelets** up to 12 mm long (awns excluded), with three awns of which the central one is plumose. **Leaf blade** up to 1 mm wide, rolled, rough, curled, leaves mainly basal. **Leaf sheath** generally glabrous, never densely hairy. **Ligule** a ring of short hairs.

Habitat Mainly in dry parts, on loose sandy soil. **Biomes**: Savanna, Nama-Karoo, Succulent Karoo.

General A palatable grass, regarded as one of the most valuable grazing grasses in its distribution range. Also regarded as a good sand binder. Often occurs in association with *Stipagrostis ciliata* var. *capensis* which has a ring of white hairs around the nodes and is a taller plant (up to 1 m) than *Stipagrostis obtusa*. **Grazing value** very high. **Ecological status**: Decreaser.

STIPAGROSTIS ZEYHERI subsp. SERICANS

Cape Bushman Grass
Drieveerboesmangras

A tufted perennial with erect culms, up to 0,75 m tall. **Inflorescence** a narrow contracted panicle, up to 200 mm long. Flowers from January to May. **Spikelets** up to 14 mm long (awns excluded), all three awns shortly plumose, except at the tips, central awn up to 24 mm long, with the other two slightly shorter. **Leaf blade** rolled.

Habitat Generally on rocky outcrops, stony slopes and in disturbed areas such as uncultivated lands. Usually grows on sandy soil. **Biomes**: Grassland and Savanna.

General Grazing value unknown, but the leaves are probably too hard to be of any grazing value.

TRIRAPHIS ANDROPOGONOIDES

Triraphis
Perdegras

A tufted perennial with a creeping rhizome and culms up to 1,2 m tall. **Inflorescence** a dense contracted or open panicle, up to 300 mm long. Flowers from October to May. **Spikelets** up to 10 mm long, with the central awn of the lemma shorter than the lemma. **Leaf blade** up to 6 mm wide, usually rolled, sometimes flattened. **Ligule** a distinct ring of hairs.

Habitat Generally in open grassland and on stony slopes. Grows on good, well drained soils. **Biomes**: Grassland, Savanna, Nama-Karoo and Fynbos.

General An unpalatable species with a relatively low leaf production, rarely an important component of natural veld. Could be confused with *Triraphis schinzii* which has a lemma awn which is longer than the lemma, and a rhizome shorter than that of *Triraphis andropogonoides*. **Grazing value** low. **Ecological status**: Variable.

TRISTACHYA LEUCOTHRIX

Hairy Trident Grass
Harige-drieblomgras

A tufted perennial often with a short rhizome and culms up to 0,9 m long. **Inflorescence** a panicle consisting of up to 6 groups of spikelets, each group with three hairy spikelets. Flowers from October to March. **Spikelets** up to 45 mm long, each spikelet with a long twisted awn up to 100 mm long. **Leaf blade** up to 6 mm wide, flattened, hairy and curled when old. **Leaf sheath:** basal sheaths covered with golden brown hairs and shiny brown inside. **Ligule** a ring of hairs.

Habitat Commonly in high-rainfall sourveld, where it prefers underutilized veld and veld that is rarely burnt. Often also found on stony slopes and in marshy areas. Prefers sandy soils. **Biomes**: Grassland, Savanna and Fynbos.

General The palatability of the grass can vary considerably from region to region. It has a good leaf production and is generally well utilized in the young stage. Often shares the habitat with *Themeda triandra* or *Alloteropsis semialata* subsp. *eckloniana*. Preferred by sheep. **Grazing value** variable. **Ecological status**: mostly Increaser I.

TRISTACHYA REHMANNII

Broom Trident Grass
Besemdrieblomgras

A tufted perennial with unbranched culms up to 0,9 m tall, with one node per culm, nodes hairy. **Inflorescence** a panicle up to 170 mm long, consisting of spikelets arranged in groups of three at the tips of the lax branches. Flowers from November to March. **Spikelets** up to 30 mm long (awn excluded), usually glabrous, with a conspicuous awn up to 100 mm long, and two shorter awns up to 22 mm long. **Leaf blade** up to 3 mm wide, flat, glabrous, or with short hairs. **Leaf sheath** round. **Ligule** a ring of hairs.

Habitat Occurs in grassland and open bushveld. Prefers stony slopes in undisturbed veld. Sometimes found on moist soils near vleis, but mostly on shallow stony soils. **Biomes**: Grassland and Savanna.

General A hard unpalatable grass, poorly utilized by livestock for most of the season. Commonly used for broom-making. *Tristachya rehmannii* could be confused with *Themeda triandra*, but the latter can easily be distinguished by the presence of leafy inflorescence spathes.

AGROSTIS ERIANTHA var. ERIANTHA

Large Panicle Agrostis
Grootpluim-agrostis

A tufted perennial with a rhizome, culms up tot 0,7 m tall. **Inflorescence** a large open panicle, up to 300 mm long, with widely spreading, straight, rigid branches. Flowers from January to April. **Spikelets** up to 5 mm long, with lemmas bearing delicate awns, glumes up to 4,5 mm long. **Leaf blade** up to 2 mm wide and usually folded.

Habitat Grows in wet areas in open sour grassland. Sometimes a weed in cultivated lands. **Biomes**: Grassland and Savanna.

General Of little grazing value owing to a low leaf production. Could be confused with *Agrostis barbuligera*, but the latter has flattened leaves, a smaller panicle (up to 250 mm long) and no rhizome.

* AGROSTIS MONTEVIDENSIS

Fog Grass
Misbeltgras

A tufted annual with culms up to 0,6 m tall. **Inflorescence** a strongly spreading panicle with thin hair-shaped branches. Flowers from November to April. **Spikelets** up to 2,5 mm long, paleas absent and pedicels 20 mm and longer. **Leaf blade** up to 2 mm wide.

Habitat   Occurs in moist and disturbed areas in mountainous grassland and often along roadsides. **Biomes**: Grassland and Fynbos.

General   A relatively rare species with little grazing value because of its relatively low leaf production. Sometimes a weed in disturbed areas. Originally from South America.

ARISTIDA BIPARTITA

Rolling Grass
Grootrolgras

A tufted weak perennial to perennial with culms up to 0,65 mm tall. **Inflorescence** an open panicle up to 300 mm long, with stiff, horizontal branches. The entire panicle breaks off when mature and is dispersed by the wind. Flowers from October to May. **Spikelets** up to 20 mm long (awns included), straw-coloured to purple, with three awns. **Leaf blade** up to 2 mm wide, folded, hard and curled.

Habitat Common in moist areas such as near vleis, and also in disturbed areas such as overgrazed veld and along roadsides. Grows on all soil types. **Biomes**: Grassland and Savanna.

General An unpalatable grass with a low production. It can be regarded as an indicator of heavy overgrazed veld when abundant in natural veld. *Aristida bipartita* could be confused with *Aristida effusa* and *Aristida scabrivalvis* subsp. *scabrivalvis*, but both those grasses are loosely tufted annuals. **Grazing value** very low. **Ecological status**: Increaser IIc.

ARISTIDA CONGESTA subsp. BARBICOLLIS

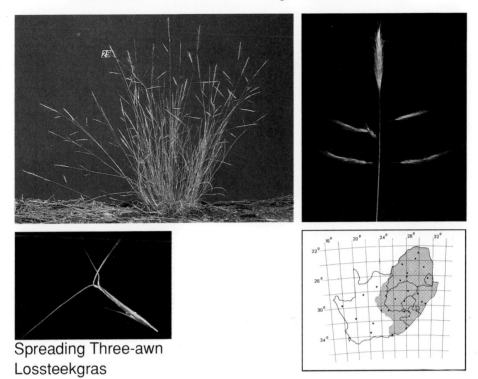

Spreading Three-awn
Lossteekgras

A tufted annual to perennial, with culms up to 0,6 m tall. **Inflorescence** an open panicle up to 200 mm long, consisting of up to 15 clusters of spikelets. Flowers from October to May. **Spikelets** each consisting of a single floret with a tripartite awn, as is typical of most species of *Aristida*. **Leaf blade** glabrous, initially flattened, but rolled as plant matures. **Leaf sheath** keeled and hairy or glabrous. **Ligule** a ring of sparse hairs.

Habitat Widespread in bushveld or grassland areas, on all soil types, with a preference for shallow, sandy soils. Common on overgrazed veld, uncultivated lands and along roadsides. **Biomes**: Savanna and Grassland.

General A poor grazing grass with low leaf production and utilized only in the young stage. Dense stands could indicate overgrazed veld. Important as a pioneer grass to cover bare patches. 'Seeds' pose a problem for wool-bearing sheep and angora goats where it penetrates the fleece, thus lowering the quality of the wool and hair. **Grazing value** very low. **Ecological status**: Increaser IIc.

ARISTIDA DIFFUSA subsp. BURKEI

Iron Grass
Ystergras

A tufted perennial with unbranched culms up to 1,0 m tall, usually with dark nodes. **Inflorescence** an open panicle, up to 300 mm long, often as wide as long. Flowers from November to April. **Spikelets** up to 40 mm long (awns included) with three awns, as typical of most species of *Aristida*. **Leaf blade** up to 2 mm wide, hard and rolled. **Leaf sheaths** glabrous.

Habitat Usually grows on slopes on shallow stony soils. Often occurs in overgrazed veld. **Biomes**: Grassland, Savanna and Nama-Karoo.

General A hard unpalatable grass, generally poorly utilized by grazers. Can be regarded as an indicator of overgrazed veld. There are two subspecies, namely subsp. *burkei* and subsp. *diffusa*. They can be distinguished by the shorter glumes (up to 12 mm) of subsp. *burkei*, while those of subsp. *diffusa* are up to 18 mm. The latter is limited to fynbos areas. *Aristida diffusa* could also be confused with *Aristida vestita* which has woolly basal leaf sheaths. **Grazing value** very low. **Ecological status**: Increaser IIc.

ARISTIDA SCABRIVALVIS subsp. SCABRIVALVIS

Purple Three-awn
Pers-steekgras

A tufted annual with culms up to 0,85 m tall and often branched. **Inflorescence** a delicate open panicle, up to 300 mm long, with spreading branches and spikelets. Flowers from January to May. **Spikelets** up to 24 mm long (awns included), with three scabrid awns. **Leaf blade** up to 3,5 mm wide, flat and slightly rough. **Leaf sheath** keeled. **Ligule** an inconspicuous membrane.

Habitat Common in disturbed areas such as uncultivated lands and roadsides, but sometimes also in undisturbed open bushveld. Grows on all soil types, but prefers shallow and sandy soils. Often found in limy soil. **Biome**: Savanna.

General Has virtually no grazing value owing to its low yield. 'Seeds' can cause problems when they penetrate the skins of sheep and other livestock. Could be confused with *Aristida bipartita* which is, however, a perennial. Can be distinguished form *Aristida scabrivalvis* subsp. *contracta* by the spikelets of the latter which are grouped in clusters at the end of the branches.

* AVENA FATUA

Common Wild Oat
Gewone Wildehawer

A tufted annual with erect culms, up to 1,0 m tall. **Inflorescence** an open panicle up to 400 mm long, consisting of drooping spikelets, hanging from slender, lax branches. Flowers from August to November. **Spikelets** consist of 2 or 3 florets up to 30 mm long, with one dark brown bent awn (up to 40 mm long) per floret. Each floret is surrounded by brown hairs. **Leaf blade** up to 10 mm wide, flattened and scabrous. **Leaf sheath** round and papery. **Ligule** a membrane up to 5 mm long.

Habitat A common weed in disturbed areas such as gardens, cultivated lands and roadsides, particularly on sandy soils. **Biomes**: Savanna and Fynbos.

General A worldwide weed originally from Europe. In South Africa especially important in the winter grain regions of the Transvaal, O.F.S. and the winter-rainfall regions. Used for haymaking in the U.S.A. *Avena fatua* can be distinguished from the closely related *Avena sterilis* which does not have an awn for every floret and can have up to 5 florets per spikelet. *Avena barbata*, on the other hand, has only two florets per spikelet, with white or cream-coloured hairs.

BOTHRIOCHLOA BLADHII

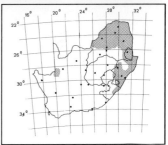

Purple Plume Grass
Persklossiegras

A tufted perennial with culms up to 1,8 m tall. **Inflorescence** a panicle up to 200 mm long, varying from notably dense to sparse, consisting of numerous racemes. Flowers from November to April. **Spikelets** occur in pairs, one spikelet pedicellate and awnless, the other sessile and bearing an awn; with pits on the lower glumes (see photograph). **Leaf blade** up to 12 mm wide, glabrous and shiny with a white midrib and a rough leaf margin. **Leaf sheath** glabrous and round. **Ligule** a short hairless membrane.

Habitat Generally in vleis, on riverbanks and other moist sites. Often along roads where water accumulates. **Biomes**: Savanna and Nama-Karoo.

General The grass has a relatively good leaf production, but is avoided by grazers because of the aromatic oils it contains. *Bothriochloa bladhii* is often confused with *Bothriochloa insculpta*, but it can be distinguished by the absence of a ring of white hairs around the nodes, which is present in the latter species. **Grazing value** low. **Ecological status**: mostly Increaser I.

BRACHIARIA DEFLEXA

False Signal Grass
Bastersinjaalgras

An open tufted annual with culms up to 0,6 m (rarely 1,0 m) tall, often branched. **Inflorescence** open, up to 200 mm long, consisting of relatively straight racemes up to 100 mm long, with sparse and shortly pedicellate spikelets. Flowers from December to June. **Spikelets** up to 3,4 mm long, usually in pairs, but sometimes solitary. **Leaf blade** up to 20 mm wide, slightly rounded at the base, hairy below and often also above. **Leaf sheath:** basal sheaths often hairy. **Ligule** a ring of short hairs.

Habitat Occurs in parts with moderate to low rainfall. Prefers shady and moist places in open bushveld. Often found in disturbed areas and is sometimes a weed in gardens. Generally grows on sandy to sandy loam soils. **Biomes**: Savanna, Grassland and Nama-Karoo.

General A palatable grass with a medium to low leaf production. Rarely occurs in thick stands in the veld (except occasionally under dense vegetation). Reasonably drought-resistant. Could be confused with certain species of *Panicum*, but the inflorescences of species of this genus are more profusely branched and the spikelets occur less in pairs.

* BRIZA MINOR

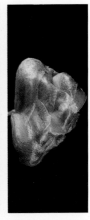

Little Quaking Grass
Kleinklokkiegras

A tufted annual with culms up to 0,7 m tall, often branched and with dark nodes. **Inflorescence** an open panicle, up to 150 mm long, consisting of spikelets borne on rigid pedicels. Flowers from September to December. **Spikelets** up to 5 mm long, glabrous, glossy and awnless. **Leaf blade** up to 9 mm wide, thin and flat. **Leaf sheath** smooth and round. **Ligule** a white membrane, up to 8 mm long.

Habitat Common in disturbed areas such as roadsides, orchards, fields under irrigation and gardens. Generally found in moist soils or in shady areas; often near rivers or vleis. Prefers loam or clay. **Biomes**: Savanna, Grassland and Fynbos.

General Originally from the Mediterranean areas, but now widespread in most countries with a moderate climate. A palatable grass, but with a low leaf production, rarely found in natural veld. Used in flower arrangements and planted in gardens, but less so than *Briza maxima*.

EHRHARTA CALYCINA

Common Ehrharta
Gewone Ehrharta

A variable tufted perennial (occasionally an annual), often with a rhizome, with culms up to 1,2 m tall. **Inflorescence** a panicle, up to 250 mm long, with a lax primary axis and drooping branches. Flowers from July to June, but usually in spring. **Spikelets** up to 8,5 mm long, awnless, with hairy lemmas and almost equal glumes. **Leaf blade** up to 7 mm wide, flattened or rolled, hairy or glabrous. **Leaf sheath** keeled, usually glabrous. **Ligule** a short membrane.

Habitat Adapted to a large variety of habitats, but usually prefers disturbed areas and sandy soils. **Biomes**: Succulent Karoo, Fynbos and Savanna.

General Regarded as good grazing and one of the few good pasture grasses in the winter-rainfall region. In Australia and California it is planted as a drought-resistant pasture. However, in certain parts of Australia it has also become a troublesome weed. *Ehrharta calycina* is a variable grass and several ecotypes and other variations can be distinguished. **Grazing value** high. **Ecological status**: Decreaser.

ERAGROSTIS ASPERA

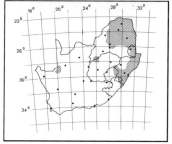

Rough Love Grass
Grootpluim-eragrostis

A tufted annual with culms up to 0,8 m tall. **Inflorescence** a large open panicle, up to 350 mm long (rarely up to 0,7 m), which can measure up to half the height of the entire plant. Flowers from February to June. **Spikelets** up to 10 mm long, pedicellate, awnless, with glumes almost of equal length. **Leaf blade** up to 10 mm wide, flat, with a distinct midrib and scattered hairs on the leaf margin and leaf base. **Leaf sheath** round, with a hairy margin. **Ligule** a ring of short white hairs.

Habitat Common in disturbed areas such as gardens and uncultivated lands. Often occurs in or near shady areas. Grows on any soil type, but prefers poor sandy soils, particularly in dolomitic areas. **Biome**: Savanna.

General A poor grazing grass, probably with a very low grazing value due to its low leaf production. A common weed in disturbed areas, which is easy to control mechanically.

ERAGROSTIS BIFLORA

Shade Eragrostis
Skadu-eragrostis

A tufted annual, with culms up to 0,7 m tall. **Inflorescence** a delicate much-branched panicle, up to 350 mm long, with long and slender pedicels. The inflorescence measures one third to half the length of the entire plant. Flowers from September to May. **Spikelets** up to 2,5 mm long, with 1 or 2 florets, from green to completely purple. **Leaf blade** up to 8 mm wide, flattened and thin. **Leaf sheath** keeled and glabrous. **Ligule** a ring of short hairs.

Habitat A shade-loving grass, which usually grows in disturbed areas, often under trees in nitrogen-rich soils. **Biomes**: Savanna, Grassland and Nama-Karoo.

General Of little value as grazing, owing to its low leaf yield. A weed in moist, shady areas such as gardens. Could be confused with *Sporobolus* and *Panicum* species, but *Eragrostis biflora* rarely has only one floret per spikelet, whereas those genera always have one floret per spikelet.

ERAGROSTIS CHLOROMELAS

Curly Leaf
Krulblaar

A tufted perennial with culms up to 0,8 m tall. **Inflorescence** a much-branched open panicle, up to 150 mm long, with lower branches usually solitary. Flowers from December to May. **Spikelets** up to 6 mm long, dark green to grey-green. **Leaf blade** thread-shaped, relatively short and typically curled. **Leaf sheath:** basal sheaths glabrous or inconspicuously hairy. **Ligule** a ring of hairs.

Habitat Generally in disturbed areas such as overgrazed or trampled veld and roadsides. Often also on undisturbed stony slopes. Grows on sandy to loamy soils. **Biomes**: Grassland and Savanna.

General A reasonably palatable grass which can endure heavy grazing. Can be used to stabilize inland quicksand. *Eragrostis chloromelas* is often confused with certain forms of *Eragrostis curvula*, but can be distinguished by the curled leaves of *Eragrostis chloromelas*. **Grazing value** average. **Ecological status**: Increaser IIb or IIc.

ERAGROSTIS CURVULA

Weeping Love Grass
Oulandsgras

A tufted perennial with culms varying in height from 0,5 m—1,5 m. **Inflorescence** a variable contracted or open panicle up to 300 mm long. Flowers from October to December. **Spikelets** up to 10 mm long, awnless, glabrous and usually dark in colour, but not exclusively so. **Leaf blade** up to 3 mm wide, hairy or glabrous, long and narrow, the colour varying from bright green to blue-green. Leaves are usually concentrated at the base of the plant. **Leaf sheath** generally distinctly veined. **Ligule** a ring of short hairs.

Habitat Common in disturbed areas on well drained, fertile soils such as uncultivated lands. Generally associated with overgrazed and trampled veld in high-rainfall areas. **Biomes**: Savanna, Grassland, Fynbos, Nama-Karoo and Succulent Karoo.

General One of the economically most important pasture grasses in the country. It is highly productive, provides early grazing in spring and establishes easily. However, palatability is medium to low. It is commonly used for pasture, hay and the maintenance of erosion control works. *Eragrostis curvula* is sometimes confused with the closely related *Eragrostis chloromelas*, but the latter can be distinguished by its curly leaves. *Eragrostis curvula* is a very variable grass. Preferred by blesbok. **Grazing value** average under natural conditions. **Ecological status**: mostly Increaser IIb.

ERAGROSTIS HETEROMERA

Bronze Love Grass
Rooikopergras

A tufted perennial with culms usually unbranched and up to 1,0 m tall. **Inflorescence** an open panicle, up to 300 mm long, consisting of lax, flexible branches, with the basal branches solitary or arranged in whorls. Flowers from January to March. **Spikelets** up to 9 mm long, glabrous, purple to green, with pedicels usually shorter or as long as the spikelets. **Leaf blade** up to 4 mm wide, flat or rolled, soft, with a very delicate tip. **Leaf sheath** keeled, with basal sheaths generally glabrous, but sometimes inconspicuously hairy. **Ligule** a ring of short hairs.

Habitat Common in grassland and bushveld, on moist sandy soils or black clay. Prefers moist sites such as the edges of seasonal pans and disturbed areas where water accumulates. **Biomes**: Grassland and Savanna.

General A reasonably palatable grass, but with a limited leaf production and probably with an average grazing value. Often confused with *Eragrostis rotifer* which has densely hairy basal leaf sheaths, as opposed to the glabrous or inconspicuously hairy basal leaf sheaths of *Eragrostis heteromera*.

ERAGROSTIS LEHMANNIANA var. LEHMANNIANA

Lehmann's Love Grass
Knietjiesgras

A tufted perennial with culms up to 0,6 m tall, often branched, lower nodes geniculate, sometimes forming roots. **Inflorescence** a slightly contracted to open panicle, up to 200 mm long. Flowers from November to June. **Spikelets** up to 8 mm long, usually dark green to greyish green, flattened and glabrous. **Leaf blade** up to 2,8 mm wide, relatively straight, with a hard tip, flattened or rolled. **Leaf sheath**: lower sheaths papery and straw-coloured. **Ligule** a ring of white hairs.

Habitat   Widespread in parts with a moderate to low rainfall. Generally in disturbed areas such as overgrazed veld, uncultivated lands and roadsides where it can form dense stands. Prefers sandy to sandy loam soils. **Biomes**: Savanna and Nama-Karoo.

General   A reasonably palatable pasture grass which is well grazed in the young stage. Able to withstand heavy grazing. Dense stands are an indicator of previously disturbed sites. A valuable pasture grass in the very dry parts of the country. Plays an important role in the stabilization of denuded and eroded soils. Is sometimes confused with the closely related *Eragrostis trichophora*, but the latter can be distinguished by the lower branches of the inflorescence arranged in a whorl as opposed to only 1 or 2 lower branches in *Eragrostis lehmanniana* var. *lehmanniana*. **Grazing value** mostly average. **Ecological status**: Increaser IIb or IIc.

ERAGROSTIS MICRANTHA

Finesse Grass
Finessegras

A weak tufted perennial with unbranched culms up to 0,9 m tall. **Inflorescence** a much-branched, lax and spreading panicle, up to 300 mm long. Flowers from January to May. **Spikelets** up to 4 mm long, with glumes unequal to almost equal, usually acute. **Leaf blade** up to 3 mm wide, with a filiform tip, flattened or rolled. **Leaf sheath** glabrous.

Habitat Generally in disturbed moist areas on sandy, loamy and limy soils, usually near vleis, pans or other wet areas. Often found in light shade. **Biomes**: Grassland, Nama-Karoo and Savanna.

General Probably of little grazing value due to its relatively low leaf production. Rarely an important component of natural veld. The inflorescence of *Eragrostis micrantha* could be confused with that of *Eragrostis chloromelas*, but the latter has a dense and strong perennial growth form and curled, filiform leaves.

ERAGROSTIS OBTUSA

Dew Grass
Douvatgras

A tufted perennial (sometimes annual), with culms up to 400 mm tall and often geniculate. **Inflorescence** an open to slightly contracted panicle, up to 120 mm long, with the lower branches usually solitary. Flowers from July to May. **Spikelets** up to 5 mm long, oval to oblong-oval and laterally compressed. **Leaf blade** up to 4,5 mm wide, usually rolled, glabrous or sparsely hairy. **Leaf sheath** usually glabrous, with a distinct collar.

Habitat Common in disturbed areas such as roadsides and overgrazed or trampled veld. Usually grows on sandy and limy soils. **Biomes**: Grassland, Savanna and Nama-Karoo.

General Well utilized, especially in the young stage, but it has a relatively low leaf production. Can be regarded as an indicator of heavily overgrazed veld. Can be distinguished from *Eragrostis echinochloidea* by the acute glumes of the latter. **Grazing value** average. **Ecological status**: Increaser IIc.

ERAGROSTIS PSEUDOSCLERANTHA

Footpath Love Grass
Voetpad-eragrostis

A weak perennial with geniculate culms up to 400 mm tall, rooting from the lower nodes and sometimes with stolons. **Inflorescence** an open panicle up to 75 mm long, with spikelets on pedicels of different lengths. Flowers form October to April. **Spikelets** up to 10 mm long, glabrous, grey to greyish green. **Leaf blade** up to 4,5 mm wide, flat, glabrous or with sparse hairs on the surface. **Leaf sheath** keeled, hairy. **Ligule** a ring of short hairs.

Habitat Generally in disturbed areas such as footpaths, overgrazed and trampled veld, uncultivated lands, bare patches and roadsides. Also found in open short grassland or under trees. Grows on most soil types, but prefers stony, sandy loam soils. **Biomes**: Savanna and Grassland.

General Probably of little value as grazing because of its low leaf production. A weed in disturbed areas, but easy to control mechanically. Could be confused with *Eragrostis cilianensis* and *Eragrostis tenuifolia*, but both those species have green spikelets, not grey as in *Eragrostis pseudosclerantha*. *Eragrostis racemosa*, another species with which it may be confused, is a strong perennial and the inflorescence is smaller and slightly contracted. Preferred by blesbok.

ERAGROSTIS RIGIDIOR

Broad-leaved Curly Leaf
Breëkrulblaar

A tufted perennial with culms up to 1,0 m tall, straight or geniculate and often branched. **Inflorescence** an open to slightly contracted panicle, up to 350 mm long, usually with the lower branches in a whorl. Flowers from October to May. **Spikelets** up to 7 mm long, glabrous and with slightly unequal glumes. **Leaf blade** up to 6 mm wide, smooth, glabrous, flattened and curling when dry. **Leaf sheath** rounded and usually glabrous, with papery basal sheaths. **Ligule** a ring of short hairs.

Habitat Common in the warm and dry regions of our country. Generally found in open spots in bushveld, or in disturbed areas such as uncultivated lands and overgrazed veld. Grows on a wide variety of soils, but prefers sandy and loam soils. **Biome**: Savanna.

General A grass of low palatability with a medium leaf production. A relatively good pasture grass in the drier parts of the country. Well utilized only in the young stage. Could be confused with *Eragrostis lehmanniana*, but the leaves of the latter do not curl and the lower branches of its inflorescence are not whorled. **Grazing value** mostly low. **Ecological status**: Increaser IIb.

ERAGROSTIS TENUIFOLIA

Elastic Love Grass
Elastiese-eragrostis

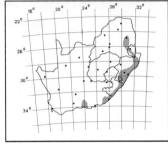

A sparse tufted annual to weak perennial, with culms up to 0,7 m tall. **Inflorescence** an open panicle, up to 300 mm long, with thin secondary branches arranged singly or in pairs on the primary axis. Flowers throughout the year, except during July and August. **Spikelets** up to 16 mm long, glossy, glabrous, toothed. **Leaf blade** up to 3 mm wide, flat, usually glabrous, except for hairs near the culm. **Leaf sheath** keeled. **Ligule** a hairy ring.

Habitat Generally in open grassland or in open patches in bushveld. Grows in disturbed areas, usually in sandy and gravelly soils, but also on clay. Often found in moist places. **Biomes**: Savanna and Grassland.

General A weed in disturbed areas and scarce in natural veld. Probably of little grazing value. *Eragrostis tenuifolia* is closely related to *Eragrostis plana* which is strongly perennial and has overlapping, fan-shaped basal leaf sheaths.

ERAGROSTIS TRICHOPHORA

Hairy Love Grass
Harige-pluimgras

A tufted perennial, sometimes forming stolons, with culms often geniculate, up to 0,6 m tall and sometimes rooting at the lower nodes. **Inflorescence** an open panicle, up to 200 mm long, consisting of a number of stiff branches arranged in whorls around the primary axis, with a ring of hairs at least at the lower whorl. Flowers from November to May. **Spikelets** up to 5 mm long, glabrous and glumes almost equal. **Leaf blade** up to 3 mm wide, flat or rolled and usually hairy. **Leaf sheath** glabrous, or with a few sparse hairs, basal sheaths papery. **Ligule** a ring of short hairs.

Habitat Occurs in the higher rainfall parts usually in disturbed areas such as roadsides and overgrazed veld. In low rainfall parts it is limited to moist soils on the edges of pans and vleis. Often occurs on shallow dolomite and limy soils. Prefers sandy to sandy loam soils. **Biomes**: Savanna and Nama-Karoo.

General A relatively palatable grass with a medium leaf production. Reasonably well grazed in the drier regions. Often confused with the closely related *Eragrostis lehmanniana* and *Eragrostis cylindriflora*. In *Eragrostis lehmanniana* the lower branches of the panicle are not whorled. *Eragrostis cylindriflora* is an annual. Preferred by goats. **Grazing value** average. **Ecological status**: mostly Increaser IIc.

ERAGROSTIS VISCOSA

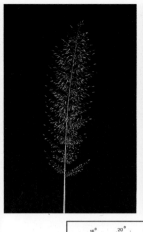

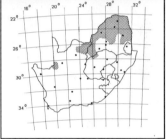

Sticky Love Grass
Klewerige-eragrostis

A tufted annual with culms up to 0,5 m tall and covered with a sticky substance at each node. **Inflorescence** a narrow open panicle with a firm, stiff primary axis, usually covered with a sticky substance. Flowers from January to August. **Spikelets** up to 3 mm long, paleas covered with long hairs and glumes almost equal in length. **Leaf blade** up to 5 mm wide, flat, mainly glabrous, with a rough leaf margin. **Leaf sheath** keeled and covered with a sticky substance at the top. **Ligule** a ring of short hairs.

Habitat Common in the bushveld parts of the country. Occurs on disturbed soils and in open sunny areas such as uncultivated lands and particularly along roadsides. Also occurs in disturbed natural veld, particularly mopani veld. Prefers dry sandy and shallow soils, but also grows in other soils. **Biome:** Savanna.

General A poor grazing grass with a low leaf production. Can be regarded as an indicator of poor soil conditions and overgrazing. May become a weed in cultivated lands, but is easily mechanically controlled. The only other grass with a sticky substance is *Eragrostis gummiflua*, but that is a perennial with glabrous spikelets.

FESTUCA SCABRA

Munnik Fescue
Munnik-swenkgras

A tufted perennial often with a rhizome, culms up to 1,0 m tall. **Inflorescence** a panicle, up to 300 mm long, usually contracted and sometimes spike-like. Flowers from September to February. **Spikelets** up to 15 mm long, with lemmas generally awnless. **Leaf blade** up to 10 mm wide, flattened or inrolled, scabrid. **Leaf sheath** velvety hairy, old sheaths splitting into fibres.

Habitat Generally in undisturbed, high-lying mountainous grassland as well as valley bushveld. Often found in moist areas and in semi-shade. Usually grows on poor sandy soil. **Biomes**: Fynbos, Savanna, Grassland and Succulent Karoo.

General An useful grass which grows in the early spring and can survive on very poor soil. *Festuca scabra* could be confused with the very closely related *Festuca caprina* and *Festuca costata*, but they have lax, open panicles and glabrous leaf sheaths.

MELINIS REPENS subsp. REPENS

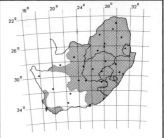

Natal Red Top
Natal-rooipluim

A tufted annual to weak perennial with culms up to 1,0 m tall, often rooting from the lower nodes. **Inflorescence** an open panicle, up to 200 mm long. Flowers from October to June. **Spikelets** up to 6 mm long, covered with reddish to pink hairs which become lighter as the plant matures. **Leaf blade** up to 11 mm wide, flattened, with a distinct midrib, hairy or glabrous. **Leaf sheath** usually hairy in the basal leaves. **Ligule** an inconspicuous ring of hairs.

Habitat Generally in disturbed areas such as uncultivated lands and roadsides. Grows on all soil types. Sometimes also found in open grassland and stony ridges. **Biomes**: Grassland, Savanna, Nama-Karoo and Fynbos.

General A fairly palatable species with a low to average production. Important as a soil stabilizer on uncultivated lands and along roadsides. Where the grass is grazed on uncultivated lands, careful management is necessary to ensure the succession of better perennial species. The inflorescence is used in flower arrangements. *Melinis repens* subsp. *repens* could be confused with *Melinis nerviglumis*, but the latter can be distinguished by its filiform leaf blades and overlapping basal leaf sheaths. Preferred by roan. **Grazing value** low. **Ecological status**: Increaser IIc.

PANICUM COLORATUM var. COLORATUM

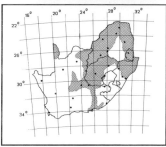

White Buffalo Grass
Witbuffelsgras

A tufted perennial with geniculate or erect culms up to 1,5 m (rarely 2 m) tall. **Inflorescence** an open panicle up to 300 mm long. Flowers from December to April. **Spikelets** up to 3 mm long, single, glabrous and awnless. **Leaf blade** up to 10 mm wide, flattened, generally hairy. **Leaf sheath** hairy or glabrous, rounded. **Ligule** a ring of short white hairs.

Habitat   Occurs in open grassland and bushveld in regions with average to low rainfall. Grows on a variety of soils, from sandy to clayey, with a preference for heavy, fertile soils that remain moist almost throughout the rainy season. **Biomes**: Savanna, Grassland and Nama-Karoo.

General   A palatable pasture grass with a high production, well utilized by grazers. Selected drought-resistant cultivars such as 'Bambatsi' and 'Bushman Mine' are used for grazing, hay and silage. Closely related to *Panicum stapfianum*. Could be confused with *Panicum maximum* which has the lower branches of the inflorescence in whorls and not solitary or in pairs as in *Panicum coloratum* var. *coloratum*. Preferred by white rhinoceros, roan, buffalo and reedbuck. **Grazing value** mostly very high.
Ecological status: Decreaser.

PANICUM DEUSTUM

Broad-leaved Panicum
Breëblaarbuffelsgras

A tufted perennial with a short rhizome and branched or unbranched culms, up to 2,0 m tall. **Inflorescence** an open panicle, up to 400 mm long, usually with solitary branching from a hairy primary axis. Flowers from September to April. **Spikelets** up to 5 mm long, glabrous and obtuse, often tinged with purple. **Leaf blade** up to 40 mm wide, flat, with a scabrid leaf margin. **Leaf sheath** round and generally hairy. **Ligule** a membrane, sometimes slightly fringed.

Habitat Occurs in open and dense bushveld and in coastal forests. Usually grows in moist low-lying areas such as riverbanks, under trees and in other shady places, and often also on stony slopes. Grows on most soil types, but prefers sandy and alluvial soils. **Biomes**: Savanna and Forest.

General A palatable and nutritious pasture grass with a relatively high leaf production. Particularly suitable as winter grazing since the grass retains most of its nutritive value and acceptability to animals during this season. Sometimes planted as a perennial for pasture. Used for the plaiting of grass mats. *Panicum deustum* could be confused with *Panicum maximum*. However, the inflorescence of *Panicum maximum* is more branched, the lower branches are arranged in whorls and the primary axis is glabrous. **Grazing value** high. **Ecological status**: Decreaser.

PANICUM DREGEANUM

Plum Panicum
Persbuffelsgras

A tufted perennial with culms up to 1,1 m tall. **Inflorescence** an open or slightly contracted panicle, up to 200 mm long, with the lower branch borne relatively low on the primary axis. Flowers from November to April. **Spikelets** up to 2,5 mm long, often purple, glabrous and acute. **Leaf blade** up to 3 mm wide, often softly hairy with a sharp tip, and usually borne basally. **Leaf sheath:** basal sheaths with fine downy hairs. **Ligule** an inconspicuous membrane.

Habitat   Occurs in regions with moderate to high rainfall. Grows in moist soils such as in vleis. Occasionally found on slopes in high-rainfall areas. Prefers alluvial and sandy soils. **Biomes**: Grassland and Savanna.

General   An attractive grass with a reasonably high leaf production and grazing value, which is sometimes well utilized by stock. Has potential as an ornamental grass, especially in grass gardens, because of its bright green leaves and purple inflorescences.

PANICUM MAXIMUM

Guinea Grass
Gewone Buffelsgras

A tufted perennial, sometimes with a short rhizome, culms up to 2,5 m tall, occasionally rooting at the lower nodes. **Inflorescence** an open panicle up to 400 mm long, with particularly the lower branches arranged in a whorl. Flowers from November to July. **Spikelets** up to 4 mm long, glabrous or hairy, often tinged with purple or entirely purple. **Leaf blade** up to 30 mm wide, flattened, glabrous or hairy, especially at ligule. **Leaf sheath** often densely hairy. **Ligule** an inconspicuous, short membrane.

Habitat Prefers damp places with fertile soil, such as in the shade of trees and shrubs and along rivers. Also adapts to a variety of other soils and conditions. **Biomes**: Savanna, Nama-Karoo and Fynbos.

General A very palatable and valuable pasture grass. Selected cultivars such as 'Green panic' produce hay and standing hay of high quality. One of the best artificial pastures for tropical and subtropical areas, responding well to fertilizers. However, if overgrazed during the summer months, the grass loses its vigour. Could be confused particularly with *Panicum deustum* and *Panicum coloratum* var. *coloratum*, but *Panicum maximum* can be distinguished by the whorled arrangement of the lower branches of the inflorescence while in *Panicum deustum* and *Panicum coloratum* var. *coloratum* these occur singly or in pairs. Preferred by most game species. **Grazing value** very high. **Ecological status**: Decreaser.

PANICUM NATALENSE

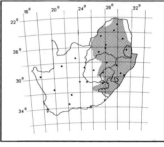

Natal Panicum
Suurbuffelsgras

A tufted perennial grass, with mostly unbranched culms up to 0,8 m tall, generally with a short rhizome. **Inflorescence** an open to slightly contracted, sometimes almost round panicle, up to 150 mm long. Flowers from October to April. **Spikelets** up to 2,5 mm long, glabrous and almost round. **Leaf blade** up to 3,5 mm wide, mostly situated at the base, usually thread-like and rolled but sometimes flat and open, with a scabrid margin. **Leaf sheath** small and round. **Ligule** inconspicuous.

Habitat Common in open sour and mountainous grassland in cool high-rainfall areas. Grows on well-drained sand or sandy loam and on stony soils, often on stony slopes or veld that is regularly burnt. **Biomes**: Grassland and Savanna.

General An unpalatable grass with hard wiry or flat leaves, which is poorly utilized by grazers. Utilized only early in spring after a veld fire. Preferred by oribi. **Grazing value** low. **Ecological status**: Decreaser.

PANICUM SCHINZII

Sweet Buffalo Grass
Soetbufffelsgras

A tufted annual with culms up to 0,9 m tall, often geniculate and branched, and often rooting at the lower nodes. **Inflorescence** an open panicle, up to 400 mm long. Flowers from November to May. **Spikelets** up to 2,5 mm long, light green to purple, pedicellate, with a blunt tip. **Leaf blade** up to 20 mm wide, dark green, glabrous, with lobes at the base. **Leaf sheath** glabrous, often with purple hues. **Ligule** an inconspicuous ring of hairs.

Habitat A moisture-loving grass, under normal conditions always near streams or vleis. A weed in cultivated lands, particularly where water accumulates. Grows on sandy or clayey soils. **Biomes**: Grassland, Savanna, Fynbos and Nama-Karoo.

General A palatable grass with a relatively high leaf production, well utilized by stock. It is also good for making hay and silage. An important weed, particularly in the summer grain production areas.

PANICUM VOLUTANS

Tumble Weed
Rolgras

A loosely tufted annual with culms up to 0,7 m tall, sometimes rooting from lower nodes. **Inflorescence** a large open panicle, up to 400 mm long, with widely spreading, stiff branches. The entire inflorescence breaks off when mature. Flowers from January to March. **Spikelets** up to 6,5 mm long, conspicuously veined, glabrous. **Leaf blade** up to 10 mm wide, flattened, hairy.

Habitat   Generally in disturbed areas such as cultivated lands and roadsides. Grows in moist areas, on black turf soils. **Biomes**: Savanna and Grassland.

General   Has a low leaf production and is therefore probably of little grazing value. May become a weed in lands of black turf soil. It is the only species of *Panicum* of which the entire inflorescence breaks off at maturity to be dispersed by the wind.

* POA ANNUA

Annual Blue Grass
Eenjarige Blougras

A tufted annual, sometimes biennial, with culms up to 200 mm (rarely up to 500 mm) tall, often rooting from the lower nodes. **Inflorescence** an open panicle, up to 120 mm long, consisting of solitary (sometimes paired) branches. Flowers throughout the year. **Spikelets** up to 6 mm long, awnless, green, often tinged with purple or white. **Leaf blade** up to 5 mm wide, lax, glabrous, with a typical blunt tip. **Leaf sheath** keeled, compressed and smooth. **Ligule** a white membrane.

Habitat Usually occurs in wet, previously disturbed areas such as gardens, lawns and roadsides. Often grows where water is available during winter, i.e. near leaking taps and water pipes. Usually found in the shade of trees and buildings. **Biomes**: Grassland, Savanna and Fynbos.

General Indigenous to Europe and Asia. A worldwide weed which can become a problem, especially in lawns. Could occur as a weed in winter-planted pastures under irrigation, but it is a weak competitor and therefore not a serious threat. Easily controlled mechanically in gardens. **Grazing value** very low. **Ecological status**: Increaser IIc.

SETARIA SAGITTIFOLIA

Arrow Grass
Pylblaargras

A loosely tufted annual with slender culms up to 0,8 m tall. **Inflorescence** an open panicle, up to 100 mm long, consisting of unilateral branches which are arranged horizontally in whorls on the primary axis. Flowers from January to March. **Spikelets** up to 2 mm long, borne on short pedicels on one side of the rachis, each with a single bristle. **Leaf blade** up to 11 mm wide, (occasionally up to 18 mm), very thin and sagittate, two acute lobes on either side of the typically short leaf stalk. **Leaf sheath**: basal sheaths slightly compressed and often finely hairy. **Ligula** a thin white membrane.

Habitat Prefers shady areas. Generally in warm localities. Grows in open to dense bushveld and in forest, on most soil types, but prefers sandy soils. **Biomes**: Savanna and Forest.

General A palatable grass, well utilized in the growing season. However, rarely an important component of natural veld. Easily distinguished by way of its arrow-shaped leaves.

SORGHUM BICOLOR subsp. ARUNDINACEUM

Common Wild Sorghum
Gewone Wildesorghum

A tufted annual to weak perennial without a rhizome and with culms up to 2,5 m tall. **Inflorescence** an open to slightly contracted panicle, up to 400 mm long, consisting of lax branches, arranged in whorls around the primary axis. Flowers from January to June. **Spikelets** borne in pairs, the one member sessile, up to 7 mm long, with a 13 mm long awn, and the other pedicellate and awnless. **Leaf blade** up to 30 mm wide, flat, with a distinct pale midrib. **Leaf sheath** round and glabrous. **Ligule** a membrane with a hairy margin.

Habitat Especially partial to wet areas such as riverbanks and vleis. Often a weed in disturbed areas such as water courses along roads and on the edge of cultivated land. Usually grows on clay but also on moist sandy soils. **Biomes**: Savanna, Grassland and Nama-Karoo.

General A good and palatable fodder grass, which is especially suitable for hay-making. However, like all *Sorghum* species, it may cause prussic acid poisoning. A problem weed in the eastern parts of the country in sugar cane fields and tropical fruit orchards. Easily confused with *Sorghum halepense* which has a long and robust rhizome. The closely related *Sorghum bicolor* subsp. *drummondii* has a compact contracted panicle, as opposed to the open panicle of *Sorghum bicolor* subsp. *arundinaceum*.

SPOROBOLUS FIMBRIATUS

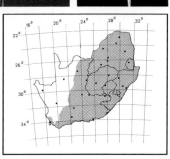

Bushveld Dropseed
Bosveldfynsaadgras

A tufted perennial with a short oblique rhizome, generally with unbranched culms up to 1,6 m tall. **Inflorescence** an open to slightly contracted panicle up to 400 mm long, consisting of numerous branches arranged irregularly or in whorls on the primary axis. Flowers from December to May. **Spikelets** up to 2 mm long, numerous, olive-green. **Leaf blade** up to 5 mm wide, glabrous, but sometimes with hairs on the leaf margin. **Leaf sheath** compressed, often with hairs on the margin. **Ligule** an inconspicuous ring of short hairs.

Habitat   Usually occurs under moist conditions such as near rivers and under trees. Often found in disturbed areas such as roadsides. Prefers well drained soils.
Biomes: Grassland, Savanna, Nama-Karoo and Fynbos.

General A palatable pasture grass with a high leaf production, well utilized by grazers. Prussic acid may occur in wilted plants. In times of famine, the 'seeds' are pounded and eaten as porridge. *Sporobolus fimbriatus* is closely related to *Sporobolus africanus*, *Sporobolus natalensis* and *Sporobolus pyramidalis* and intermediates are often formed.
Grazing value mostly very high.   **Ecological status**: Decreaser

SPOROBOLUS IOCLADOS

Pan Dropseed
Panfynsaadgras

A tufted perennial with a short rhizome, culms up to 1,0 m tall, sometimes with stolons, and rooting at the lower nodes. **Inflorescence** an open pyramid-shaped to ovate panicle, up to 200 mm long, with branches arranged in neat whorls around the primary axis. Flowers from January to April. **Spikelets** up to 2,5 mm long, glossy and glabrous, with the upper glume as long as or longer than the spikelet. **Leaf blade** up to 12 mm wide, flat, with a hairy leaf margin. **Leaf sheath** keeled and glabrous, or with short hairs, basal sheaths papery. **Ligule** a ring short hairs.

Habitat Occurs on a wide variety of soils with a preference for brackish soils. Often grows in and around seasonal pans and in disturbed areas such as roadsides. **Biomes**: Savanna, Nama-Karoo and Grassland.

General A palatable grass, well utilized by grazers, but with a low to medium leaf production. Probably an important source of food in pans in the dry parts of the country. The inflorescence may sometimes be confused with those of certain species of *Eragrostis*. The spikelets of *Sporobolus ioclados* are, however, smaller and one-flowered as opposed to the multiflowered spikelets of the *Eragrostis* species. **Grazing value** average. **Ecological status**: mostly Decreaser.

SPOROBOLUS NITENS

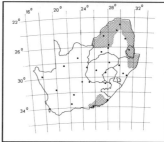

Curly-leaved Dropseed
Krulblaarfynsaadgras

A tufted perennial with a short rhizome, culms often branched, and up to 0,5 m tall, sometimes with roots at the lower nodes. **Inflorescence** an open panicle, up to 140 mm long, with the lower branches arranged in whorls. Flowers from November to April. **Spikelets** up to 1,5 mm long, borne at the tips of the branches, with the lower spikelet groups often lacking in older inflorescences. **Leaf blade** up to 8 mm wide, flattened, with a hairy or glabrous surface, leaf margin wavy and often with short, sparse hairs. **Leaf sheath** glabrous or hairy, particularly near the margin.

Habitat Generally in disturbed areas such as bare patches, overgrazed veld, and sometimes in gardens. Grows on sandy and brackish soils. **Biomes**: Savanna and Grassland.

General Probably palatable, but has a relatively low leaf production. Grows on bare patches, where few other perennial grasses can survive. *Sporobolus nitens* is sometimes confused with *Sporobolus ludwigii* which has larger spikelets (up to 2 mm long), and *Sporobolus coromandelianus* which is an annual. **Grazing value** low. **Ecological status**: Increaser IIc.

SPOROBOLUS PECTINATUS

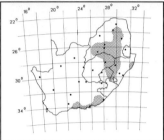

Fringed Dropseed
Kammetjiesgras

A tufted perennial with a rhizome, culms up to 0,74 m tall, usually unbranched. **Inflorescence** an open panicle, up to 200 mm long, with the branches arranged in whorls on the primary axis. Flowers from November to March. **Spikelets** up to 3,5 mm long, densely grouped at the tip of the branches. **Leaf blade** up to 8 mm wide, flattened or folded, glabrous, except for sparse hairs (resembling a comb). **Leaf sheath**: old basal sheaths papery, hairy or glabrous.

Habitat Generally in open sour grassland, particularly against stony slopes, often found on outcrops and quartzite ridges. Prefers shallow stony soil. **Biome**: Grassland.

General As a sourveld species, the grass is probably unpalatable, but it has a high leaf production. *Sporobolus pectinatus* is closely related to *Sporobolus centrifugus*, and the inflorescences in particular are easily confused, although the leaves of the latter are considerably narrower (up to 1,5 mm wide) and the sparse hairs on the margin are lacking.

SPOROBOLUS STAPFIANUS

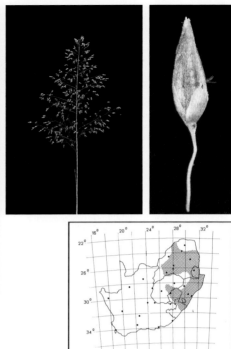

Fibrous Dropseed
Veselfynsaadgras

A tufted perennial with culms up to 0,55 m tall. **Inflorescence** an oval, open panicle with dichotomous branching, branching point on primary axis glabrous. Flowers from November to March. **Spikelets** up to 2 mm long, glossy and glabrous. **Leaf blade** up to 2 mm wide, glabrous and initially flat, but later involute. **Leaf sheath**: basal sheaths break up into fibres, with curly, woolly hairs between the fibres. **Ligule** an inconspicuous ring of hairs.

Habitat Occurs on outcrops or stony slopes in open grassland or bushveld. Often found near rivers or streams. Grows in shallow, sandy, compacted or well drained soils. **Biomes**: Grassland, Savanna and Nama-Karoo.

General A palatable grass, but with a low leaf production and thus of little grazing value. It is, however, rarely dominant in natural veld. Is sometimes confused with the closely related *Sporobolus festivus*, but *Sporobolus stapfianus* can be distinguished by the woolly hairs between the fibres of its basal sheaths, as opposed to the glabrous fibres of *Sporobolus festivus*.

STIPAGROSTIS UNIPLUMIS var. UNIPLUMIS

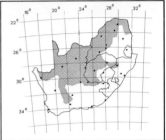

Silky Bushman Grass
Blinkblaarboesmangras

A tufted perennial with culms up to 0,9 m tall. **Inflorescence** an open to slightly contracted panicle, up to 300 mm long. Flowers from December to May. **Spikelets** up to 10 mm long (excluding awns), with three awns, the central one plumose some distance from its base, with a tuft of hair at the branching point of the awns. **Leaf blade** up to 2 mm wide, usually rolled, glabrous or sparsely hairy. **Leaf sheath** round. **Ligule** a ring of long white hairs.

Habitat Usually on undisturbed sandy soil and sometimes on flood plains. Also in disturbed places, along roads for example. **Biomes**: Savanna and Namma-Karoo.

General A fairly valuable pasture grass of variable palatability, well utilized in the young stage. Could be confused and sometimes interbreed with *Stipagrostis uniplumis* var. *neesii*, but the latter is distinguished by its inflorescence with relatively few spikelets and its glumes which are shorter than 10 mm. **Grazing value** average to high.

251

TRICHOLAENA MONACHNE

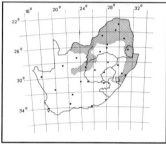

Blue-seed Grass
Blousaadgras

A tufted perennial with culms geniculate and up to 1,4 m tall, occasionally rooting from the lower nodes. **Inflorescence** an open panicle, up to 170 mm long, branches often arranged in pairs on the primary axis. Flowers from November to March. **Spikelets** up to 3 mm long, usually glabrous. **Leaf blade** up to 8 mm wide, generally rolled and glabrous (rarely with short hairs). **Leaf sheath** round. **Ligule** a ring of short white hairs.

Habitat Common in disturbed areas such as roadsides and uncultivated lands in dry bushveld regions and sometimes in open grassland. Grows on most types of soil, with a preference for sandy soils. **Biomes**: Savanna and Grassland.

General A species with an average palatability and a low leaf production. Can be an important pasture grass where dominant. Could be regarded as an indicator of disturbed areas. The inflorescence is sometimes used in flower arrangements. *Tricholaena monachne*, which is often confused with *Panicum* species, can be identified by the long pedicels on which its spikelets are borne, while *Panicum* spesies can be distinguished by its distinct lower glumes. **Grazing value** mostly average. **Ecological status**: Increaser IIb or IIc.

TRICHONEURA GRANDIGLUMIS

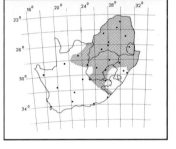

Small Rolling Grass
Klein Rolgras

A tufted perennial with culms up to 400 mm tall. **Inflorescence** very open, up to 250 mm long and 300 mm wide, consisting of 10—20 racemes borne along a firm, hairy primary axis. The entire inflorescence breaks off when mature and is then distributed by the wind. Flowers from December to January. **Spikelets** up to 14 mm long and not touching. **Leaf blade** up to 7 mm wide, flattened even when old, usually hairy. **Leaf sheath** rounded and hairy. **Ligule** a white membrane, longer than 1,5 mm.

Habitat Widespread but usually in open grassland, often found on disturbed or over-grazed sandy soils. **Biomes**: Grassland and Savanna.

General An unpalatable grass with a low leaf production. Apparently somewhat poisonous in Lesotho. Sometimes an indicator of bad veld management. Preferred by roan. **Grazing value** low. **Ecological status**: Increaser IIc.

ARISTIDA MERIDIONALIS

Giant Three-awn
Langbeensteekgras

A tufted perennial with unbranched culms up to 2,0 m tall. **Inflorescence** an open panicle up to 0,6 m long (rarely longer), with numerous pendulous, scattered spikelets. Flowers from November to May. **Spikelets** up to 50 mm long, with a tripartite awn, twisted column and borne on long thin pedicels. **Leaf blade** up to 5 mm wide and inrolled. **Leaf sheath** hard, shiny and without hairs. **Ligule** a ring of long woolly hairs.

Habitat Common in Kalahari thornveld where it grows in deep sand. Also in other parts in sand and stony soil. Often along roads and in other moist places, including vleis. **Biome**: Savanna.

General A hard, fibrous grass utilized only in the young stage. Sometimes used as thatching grass and for broom-making, also in flower arrangements. Could be confused with *Aristida vestita*, but the more hairy and woolly culms of the latter are characteristic. **Grazing value** low. **Ecological status**: Increaser IIb.

ARISTIDA SCIURUS

Tall Three-awn
Langsteekgras

A tufted perennial with a short rhizome and erect, unbranched culms up to 1,4 m tall. **Inflorescence** a dense, sometimes spike-like, hanging panicle up to 550 mm long. Flowers from January to May. **Spikelets** up to 30 mm long (awns included), with three awns and with the lower glume half the length of the upper glume, lemma not articulated. **Leaf blade** up to 3 mm wide (rarely up to 6 mm), narrower than the leaf sheaths. **Leaf sheaths:** basal sheaths often covered with woolly white hairs, particularly at the mouth of the sheath.

Habitat   Occurs mainly in mountainous sour grassland, usually on moist sandy soils and often along roadsides. **Biome:** Grassland.

General   Although the grass has a reasonable leaf production, it soon becomes hard and fibrous and unacceptable as grazing. Could be confused with *Aristida pilgeri* and *Aristida spectabilis*. *Aristida spectabilis*, however, has a membranous lower glume and *Aristida pilgeri* has a lemma articulation.

* BRIZA MAXIMA

Big Quaking Grass
Grootklokkiegras

A tufted annual with culms up to 0,6 m tall. **Inflorescence** an open panicle up to 100 mm long, with 2—12 pendulous spikelets. Flowers from July to December. **Spikelets** up to 25 mm long, glossy, awnless, with thin pedicels. **Leaf blade** up to 8 mm wide, flattened, glabrous, with a rough leaf margin. **Leaf sheath** round. **Ligule** a white membrane up to 5 mm long.

Habitat Occurs in disturbed areas such as gardens, roadsides, orchards and along irrigated fields. Often planted in gardens. Prefers well drained soil. **Biomes**: Fynbos, Savanna and Grassland.

General Originated in the Mediterranean areas, but now widespread in most countries with a temperate climate. Can be palatable, but has a very low leaf yield and is therefore of little grazing value. An attractive grass which is often used in flower arrangements. Sometimes a weed in irrigated lands.

* BROMUS CATHARTICUS

Rescue Grass
Reddingsgras

An annual to weak perennial tufted grass with unbranched culms up to 1,0 m tall. **Inflorescence** an open, drooping panicle, up to 350 mm long. Flowers from October to April. **Spikelets** up to 30 mm long, distinctly flattened, with pedicels longer than spikelets. **Leaf blade** up to 12 mm wide, hairy or glabrous, flattened and sharply keeled. **Leaf sheath** keeled. **Ligule** a membrane up to 6 mm long.

Habitat Occurs, especially as a weed, in disturbed areas such as gardens, cultivated lands and roadsides. Prefers moist patches and is often found in shade. Under normal conditions, it grows near water, on riverbanks for example. Occurs on all types of fertile soil. **Biomes:** Grassland, Fynbos and Nama-Karoo.

General Different varieties were introduced from Europe, Australia and America. A palatable grass, planted as an annual pasture or in combination with clover. Particularly suitable as a winter feed for dairy cattle. Palatability decreases during summer. A troublesome weed in lucerne lands, orchards, vineyards and other areas under irrigation. In certain regions an important stabilizer of erodible soils. **Grazing value** average.

*BROMUS DIANDRUS

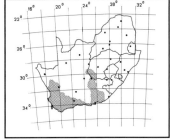

Ripgut Brome
Predikantsluis

A tufted annual grass with culms up to 0,8 m tall. **Inflorescence** an open, drooping panicle up to 300 mm long. Flowers from September to January. **Spikelets** up to 90 mm long (including awns), with 3—5 florets, spikelets attached to long, slender branches. **Leaf blade** up to 8 mm wide, flattened, with short delicate hairs on both sides. **Leaf sheath** round, with fine hairs. **Ligule** a jagged membrane.

Habitat Common in disturbed areas such as gardens, cultivated lands and roadsides. Under natural conditions limited to moist sites such as riverbanks. Widespread, but particularly common in the south-western Cape Province. **Biomes:** Nama-Karoo and Fynbos.

General A weed of economic significance in the grain producing regions of the south-western Cape. Originally from Europe and the Mediterranean areas, but now widespread throughout southern Africa. Poorly utilized by livestock owing to the long awns which cling to the mouth and nostrils. Also clings to sheep's wool, thus lowering its quality. **Grazing value** low.

* ERAGROSTIS TEF

Teff
Tef

A tufted annual with culms up to 1,0 m tall. **Inflorescence** a panicle up to 350 mm long with long flexuous branches. Flowers from November to May. **Spikelets** up to 9 mm long, glabrous and slightly glossy, generally brown when mature. **Leaf blade** up to 4 mm wide, relatively thin, soft and flattened. **Leaf sheath** usually rounded. **Ligule** a ring of hairs.

Habitat Occurs on most types of soil where veld has been disturbed, in areas with an annual rainfall of 600 mm or more. Often planted along new roads. **Biomes**: Savanna, Grassland, Fynbos and Nama-Karoo.

General Originally from north eastern Africa. A fast-growing, palatable grass, commonly planted for hay and particularly suitable for horse-fodder. Could become a weed problem in cultivated lands and gardens. Seldom occurs in natural veld. Also used for maintaining soil erosion works. An important cereal in Ethiopia. **Grazing value** high.

* PASPALUM URVILLEI

Giant Paspalum
Langbeen-paspalum

A tufted perennial with erect culms, up to 2,5 m tall. **Inflorescence** consists of 10—20 racemes arranged along the primary axis. Flowers from October to April. **Spikelets** up to 3 mm long, covered with delicate white hairs. **Leaf blade** up to 12 mm wide, flattened, with a scabrid leaf margin. **Leaf sheath** of basal leaves densely hairy, resembling prickly-pear thorns. **Ligule** a membrane with hairs along the sides.

Habitat Prefers disturbed wet conditions such as vleis, riverbanks, waterways and other places where water accumulates. Occurs on loam to clay-loam soils. **Biomes**: Grassland, Savanna and Fynbos.

General Introduced from South America as an artificial pasture grass. Palatable, with a high leaf production under ideal conditions. Palatability decreases as the plant matures. Particularly suitable for hay-making. Can continue growing throughout the winter in regions where winters are mild. *Paspalum urvillei* could be confused with *Paspalum dilatatum*, but the latter seldom has erect culms that are longer than 1,8 m. **Grazing value** average. **Ecological status**: Decreaser or Increaser I.

SETARIA MEGAPHYLLA

Broad-leaved Bristle Grass
Riffelblaarmannagras

A tufted perennial, often with a short rhizome, with culms up to 2 m tall (rarely up to 3 m). **Inflorescence** an open or contracted panicle, up to 0,6 m long, but occasionally longer. Flowers from September to June. **Spikelets** up to 3 mm long and glabrous, interspersed with scattered green bristles. **Leaf blade** up to 100 mm wide, lanceolate, conspicuously pleated lengthwise, with a scabrid leaf margin. **Leaf sheath** usually hairy. **Ligule** a ring of hairs.

Habitat A shade-loving grass, found in and around forests and plantations or in low-lying areas along rivers. Usually occurs in moist soils and often in disturbed areas such as watercourses along roads. **Biomes**: Savanna, Grassland and Forest.

General A palatable grass with a high leaf production, remaining green until late in the season. Old material becomes hard and unacceptable as grazing. However, the grass is rarely abundant in natural veld. Planted as an ornamental in gardens. Closely related to *Setaria lindenbergiana* which has smaller leaves (up 10 mm wide) and smaller inflorescences.

SORGHUM VERSICOLOR

Black-seed Wild Sorghum
Swartsaadwildesorghum

A tufted annual or weak perennial with culms up to 1,2 m tall and nodes with a conspicuous ring of white hairs. **Inflorescence** a lax, drooping panicle with thin branches usually arranged in whorls around the primary axis. Flowers from December to May. **Spikelets** occur in pairs of which one member is sessile, up to 7 mm long, dark brown or black, with a contorted awn up to 40 mm long, and the other pedicellate, green and awnless. **Leaf blade** up to 8 mm wide, usually with short hairs and a distinct midrib. **Leaf sheath** often purple, smooth. **Ligule** a short membrane.

Habitat Usually occurs in disturbed areas such as roadsides and along cultivated lands where it can form thick stands. Prefers growing on black turf. **Biomes**: Savanna and Grassland.

General A palatable grass, well utilized by livestock, but with a relatively low leaf production. As with most species of *Sorghum*, there is the danger of prussic acid poisoning with stock, especially if the grass is cut for fodder. It has medicinal uses among the Sotho and Zulu. **Grazing value** average.

TRICHOPTERYX DREGEANA

Carpet Grass
Matgras

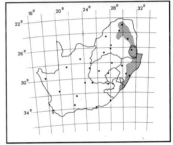

A trailing perennial with a creeping rhizome and culms up to 0,9 m tall. **Inflorescence** an open or contracted drooping panicle, up to 140 mm long (usually shorter), with filiform branches. Flowers from December to May. **Spikelets** up to 7 mm long, with a whorl of hairs at the base, the central awn up to 7 mm long and the others considerably shorter. **Leaf blade** up to 4 mm wide, directed horizontally or downwards, generally glabrous and scabrous. **Ligule** a ring of both short and long hairs.

Habitat Usually occurs in wet areas and in shade or open grassland. Often in crevices, or between rocks and on slopes. **Biomes**: Savanna and Grassland.

General Sometimes forms a dense, untidy, tangled mat which can play an important role in erosion control, particularly on steep slopes. Could be confused with *Eragrostis volkensii* which has the same growth form and also occurs in wet areas. However, the leaves of *Eragrostis volkensii* are wider (up to 8 mm) and blue-green as opposed to the light green leaves of *Trichopteryx dregeana*.

ANDROPOGON EUCOMUS

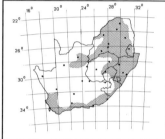

Snowflake Grass
Kleinwitbaardgras

A tufted perennial with culms up to 0,9 m tall. **Inflorescence** a false panicle consisting of 2—6 groups. Each group has 3—5 digitate racemes covered with silver white hairs and borne terminally on the branches. Flowers from November to May. **Spikelets** 2—3 mm long, hairy, with a single awn. **Leaf blade** keeled and glabrous, except for fine hairs near ligule. **Leaf sheath** distinctly flattened and glabrous. **Ligule** an inconspicuous membrane with sparse, scattered hairs.

Habitat Occurs mostly on disturbed soils where water accumulates, for example along roadsides and on uncultivated lands, or otherwise grows on acid, sandy soils or in undisturbed sandveld vleis. **Biomes:** Fynbos, Savanna and Grassland.

General Indicates seepage areas and poorly drained soil. The inflorescence is sometimes used in flower arrangements. Could be confused with *Andropogon huillensis* which is a larger plant (up to 1,8 m tall) with bigger spikelets (up to 5 mm long). **Grazing value** very low.

ANDROPOGON GAYANUS var. POLYCLADUS

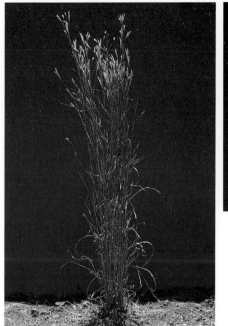

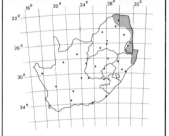

Blue Grass
Blougras

A tufted perennial with erect culms up to 2,5 m tall. **Inflorescence** a false panicle consisting of paired racemes up to 100 mm long. Flowers from December to May. **Spikelets** generally glabrous, appear in pairs, one spikelet being sessile with an awn up to 30 mm long, and the other pedicellate with an awn up to 10 mm long. **Leaf blade** up to 25 mm wide, very narrow at the base with a prominent midrib. **Leaf sheath** rounded. **Ligule** an inconspicuous round membrane with a fringed margin.

Habitat Grows on most soils, often on soil with a high magnesium content, showing a preference for shady spots and stony places. **Biome:** Savanna.

General A good, palatable pasture grass which is well utilized in the young stage. The grass is, however, seldom an important component of natural veld. A relatively drought-resistant grass, occasionally used as thatching grass where it is abundant. **Grazing value** probably average.

ANDROPOGON HUILLENSIS

Large Silver Andropogon
Grootwitbaardgras

A tufted perennial, usually with a short rhizome and culms up to 1,8 m tall. **Inflorescence** consists of 5—7 flowering branches, each bearing 4—10 feathery racemes of different lengths (up to 100 mm) which are arranged digitately or semi-digitately at the tip of the branch. Flowers from September to June. **Spikelets** up to 5 mm long, sessile and glabrous, with a thin awn and an additional hairy stalk at the base which may bear a spikelet. **Leaf blade** up to 4 mm wide, with a distinct white midrib, upper leaves generally flat and lower leaves folded. **Leaf sheath**: basal sheaths folded. **Ligule** a short white membrane.

Habitat Grows in wet areas such as riverbanks and edges of vleis, in open grassland or bushveld. Sometimes al so occurs in watercourses along roads. Prefers wet sandy soil. **Biomes:** Savanna and Grassland.

General Possibly an average to low grazing value due to its tendency to harden as the growing season progresses. A good indicator of wet sandy soil. Sometimes confused with the closely related *Andropogon eucomus*, but the latter is much smaller (up to 0,9 m high) and occurs more abundantly.

CYMBOPOGON EXCAVATUS

Broad-leaved Turpentine Grass
Breëblaarterpentyngras

An tufted perennial with unbranched culms, up to 1,5 m tall. **Inflorescence** a thick, false panicle 60—300 mm long, consisting of numerous pairs of racemes, each partially enclosed by a spathe. Flowers from November to May. **Spikelets** occur in pairs: one spikelet pedicellate and awnless, and the other sessile, up to 5 mm long and with a twisted awn. **Leaf blade** up to 14 mm wide, glabrous and flattened, often with a wax coating and a distinct midrib. **Leaf sheath** glabrous and round. **Ligule** a short membrane.

Habitat Occurs on most soils, but prefers stony, sandy soils in disturbed as well as undisturbed sour grassveld. **Biomes:** Grassland and Savanna.

General An unpalatable grass which is poorly utilized because of a strong turpentine taste. It may be utilized on over-grazed veld or after a veld fire, where the grass tends to recover faster than others. However, it is very unpalatable and hard in its mature stage. It is sometimes used as thatching grass. **Grazing value** low. **Ecological status**: mostly Increaser I.

CYMBOPOGON PLURINODES

Narrow-leaved Turpentine Grass
Smalblaarterpentyngras

A tufted perennial with culms up to 1,0 m tall. **Inflorescence** a false panicle, up to 300 mm long, consisting of up to eight hairy, paired racemes up to 20 mm long, each with a leaf-like spathe. Flowers from October to May. **Spikelets** occur in pairs, one spikelet sessile, up to 6 mm long, with an awn, and the other pedicellate and awnless. **Leaf blade** up to 4 mm wide, glabrous and sometimes folded, with a fine, curled tip, most leaves concentrated at the base of the plant. **Leaf sheath** round, basal sheaths sometimes densely hairy. **Ligule** a membrane with a hairy margin.

Habitat Occurs on most soil types in stable veld. Often in association with *Themeda triandra*. **Biomes:** Grassveld, Savanna, Nama-Karoo, Fynbos and Succulent Karoo.

General An aromatic grass with a turpentine taste and smell, generally unpalatable. It is however, utilized in the young stage, or after a veld fire. *Cymbopogon plurinodes* can be distinguished from *Cymbopogon excavatus* by the wider leaf blades (up to 14 mm) of the latter species. Preferred by Cape mountain zebra. **Grazing value** usually low but may vary. **Ecological status**: mostly Increaser I.

CYMBOPOGON VALIDUS

Giant Turpentine Grass
Reuse Terpentyngras

A tufted perennial with culms up to 2,4 m tall. **Inflorescence** a false panicle, up to 350 mm long, consisting of groups of paired racemes, each partially enclosed by a leaf-like spathe. Flowers from July to June. **Spikelets** in pairs, one member sessile, up to 6 mm long, bearing an awn, and the other pedicellate, slightly shorter and awnless. **Leaf blade** up to 10 mm wide, glabrous, with a distinct midrib and a rough leaf margin. **Leaf sheath**: lower ones short and hairy. **Ligule** a conspicuous membrane up to 7 mm long.

Habitat Usually occurs in mountainous grassland in the high-rainfall parts of the country. Grows in vleis, wet sites, along roads and on the margin of tree communities. Prefers stony slopes with loamy soil. **Biomes:** Grassland and Savanna.

General An unpalatable grass owing to its strong, aromatic, turpentine-like taste, and thus of little grazing value. It is utilized only in the young stage, when little other grazing is available. **Grazing value** low. **Ecological status**: Increaser I.

HYPARRHENIA ANAMESA

Bundle Thatching Grass
Gerftamboekiegras

A tufted perennial with a rhizome and culms up to 1,2 m tall. **Inflorescence** a false panicle consisting of raceme pairs up to 25 mm long. Each raceme pair with 4—7 awns which are up to 40 mm long. Flowers from October to May. **Spikelets** occur in pairs of which one spikelet is sessile, up to 6,5 mm long and awned, while the other is pedicellate and usually awnless. **Leaf blade** up to 4 mm wide.

Habitat   Occurs in open dry areas, and often along roadsides. **Biomes**: Grassland, Savanna and Fynbos.

General   Probably utilized at an early stage, but becomes hard and fibrous later and thus unacceptable as grazing. A species described fairly recently and which forms an intermediate between *Hyparrhenia hirta* and *Hyparrhenia filipendula*. *Hyparrhenia hirta* has longer racemes (up to 40 mm) and 8—14 awns per raceme pair, while *Hyparrhenia filipendula* has shorter racemes (up to 12 mm) with 2—4 awns per raceme pair.

HYPARRHENIA CYMBARIA

Boat Thatching Grass
Bootjietamboekiegras

A tufted perennial with a rhizome and culms generally up to 2 m tall (rarely up to 4 m), often rooting from the lower nodes. **Inflorescence** a false panicle up to 400 mm long (rarely longer), consisting of raceme pairs up to 15 mm long, each supported by a boat-shaped, papery spathe, up to 18 mm long. Flowers from November to June. **Spikelets** occur in pairs of which one member is sessile, up to 4,5 mm long, with a thin curved awn up to 16 mm long, and the other is pedicellate and awnless. **Leaf blade** up to 20 mm wide, flat, with a distinct midrib and a scabrid leaf margin. **Leaf sheath** round. **Ligule** membranous.

Habitat Common in high-rainfall regions in grassland, forest edges or open patches in dense bushveld and often along roadsides. Usually occurs in high-lying areas, but also in low-lying parts and then particularly in shady places near streams. **Biomes**: Grassland and Savanna.

General A palatable grass, acceptable as grazing in the young stage, but soon after the flowering season it becomes hard and unpalatable. Sometimes confused with *Hyparrhenia variabilis*, but the latter has longer awns (up to 30 mm) and longer spathes (up to 25 mm).

HYPARRHENIA DICHROA

Red Thatching Grass
Rooitamboekiegras

A tufted perennial with a rhizome and strong culms up to 3,0 m tall. **Inflorescence** a false panicle up to 0,6 m long, consisting of pairs of racemes up to 30 mm long, each supported by a lanceolate spathe, each pair of racemes with 6—10 awns. Flowers from March to June. **Spikelets** occur in pairs of which one member is sessile and up to 5 mm long, and the other pedicellate. **Leaf blade** up to 8 mm wide, flattened and usually glabrous.

Habitat Generally in moist and disturbed areas and often along roadsides. **Biome**: Savanna.

General Quickly becomes hard, fibrous and unacceptable as grazing. Is a popular thatching grass. Difficult to distinguish from *Hyparrhenia rufa* and intermediates also occur. However, *Hyparrhenia rufa* usually has more awns (7—14) per raceme.

HYPARRHENIA FILIPENDULA var. PILOSA

Fine Thatching Grass
Fyntamboekiegras

A tufted perennial with culms up to 1,5 m tall (rarely up to 2,0 m). **Inflorescence** a long false panicle, consisting of a number of paired racemes, up to 12 mm long, each pair with four conspicuous awns up to 55 mm long. Flowers from November to April. **Spikelets** up to 7 mm long, of two types, one sessile with a long awn and the other pedicellate with a short, straight awn. **Leaf blade** up to 4 mm wide, flattened and with a distinct midrib. **Leaf sheath** round. **Ligule** a papery membrane.

Habitat Occurs in bushveld in high-rainfall areas. In the drier parts it is found mostly along roadsides, near vleis and rivers. Grows on all types of soil. **Biome**: Savanna.

General A grass of medium palatability, well utilized in the early growing season. From the middle of the growing season it becomes hard and woody. A very good thatching grass and a better pasture grass than most of the other thatching grasses. Sometimes confused with the closely related *Hyparrhenia hirta*, but the latter can be distinguished by the fact that it has at least eight awns per raceme pair, whereas *Hyparrhenia filipendula* has four at the most. Two varieties can be distinguished, i.e. *Hyparrhenia filipendula* var. *filipendula* which usually has two awns per raceme pair and, *Hyparrhenia filipendula* var. *pilosa* which has four awns per raceme pair. **Grazing value** average. **Ecological status**: Increaser I.

HYPARRHENIA HIRTA

Common Thatching Grass
Dekgras

A tufted perennial with culms up to 1,0 m tall, sometimes with a short rhizome. **Inflorescence** a false panicle up to 300 mm long, consisting of a number of raceme pairs, each pair with 8—14 awns up to 35 mm long, and subtended by a short spathe. Flowers from September to March. **Spikelets** occur in pairs, one member sessile, up to 6,5 mm long and awned, and the other pedicellate and awnless. **Leaf blade** up to 3 mm wide, mostly glabrous and flattened. **Leaf sheath** keeled. **Ligule** an inconspicuous brown membrane.

Habitat Common in open grassland, on rocky slopes and along rivers. Occurs on most soil types, with a preference for well drained stony soils. Forms dense stands in disturbed areas such as uncultivated lands and roadsides where it can keep out other grasses for many years. **Biomes**: Grassland, Savanna and Nama-Karoo.

General Considered to be a relatively good pasture grass, particularly early in the season before it becomes hard and fibrous. Highly resistant to cold and drought. Play an important role in stabilizing bare and sandy soils and protecting them against erosion. *Hyparrhenia hirta* is sometimes confused with the closely related *Hyparrhenia filipendula*, but the latter can be distinguished by its 2—4 awns per raceme pair. Preferred by oribi, roan and sable. **Grazing value** average. **Ecological status**: mostly Increaser I.

HYPARRHENIA TAMBA

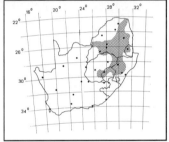

Blue Thatching Grass
Bloutamboekiegras

A tufted perennial with a rhizome, culms up to 3,0 m tall and with prop roots at the lower nodes. **Inflorescence** a false panicle, up to 0,6 m long, with paired racemes up to 20 mm long, each supported by a spathe up to 40 mm long. Flowers from December to June. **Spikelets** occur in pairs of which one member is sessile and 5 mm long with a curved awn up to 25 mm long, and the other is pedicellate with a short, straight awn. **Leaf blade** up to 7 mm wide, flat, hairy or glabrous. **Leaf sheath:** basal sheaths hairy. **Ligule** a membrane, sometimes with a hairy margin.

Habitat Occurs is grassland and bushveld areas on moist soils near streams and vleis. Common along roads. **Biomes**: Savanna and Grassland.

General As pasture it is probably too hard for most grazers. The grass is sometimes used for thatching. *Hyparrhenia tamba* may be confused with the closely related *Hyparrhenia collina* and *Hyparrhenia dregeana*, but the latter has 10—25 awns per raceme pair as opposed to the 5—8 awns of *Hyparrhenia tamba*, and *Hyparrhenia collina* is usually a smaller plant (up to 1,3 m tall) with lax culms. **Grazing value** low. **Ecological status**: Increaser I.

HYPERTHELIA DISSOLUTA

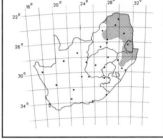

Yellow Thatching Grass
Geeltamboekiegras

A tufted perennial with culms generally unbranched and up to 3 m tall. **Inflorescence** a false panicle consisting of a number of paired racemes, each with a purple to red spathe and two yellow twisted awns. Flowers from January to June. **Spikelets** occur in pairs of which one member is up to 14 mm long and pedicellate, with a short straight awn, and the other up to 7,5 mm long and sessile, with a long twisted awn up to 90 mm long. **Leaf blade** up to 12 mm wide, smooth and flat, with a distinct midrib. **Leaf sheath** with distinct auricles. **Ligule** a membrane.

Habitat Common in disturbed areas such as uncultivated lands, and particularly along roadsides in bushveld. Grows on all types of soil. **Biome**: Savanna.

General Relatively palatable when young, but becomes woody later in the season and then poorly utilized by most grazers. Commonly used as thatching grass. Can be distinguished from most species of *Hyparrhenia* by the long, yellow, relatively thick awns and the yellow culms. **Grazing value** low to average. **Ecological status**: mostly Increaser I.

MONOCYMBIUM CERESIIFORME

Boat Grass
Bootjiegras

A tufted perennial with culms up to 0,6 m tall, occasionally with a short rhizome. **Inflorescence** an open false panicle consisting of 5—20 racemes which are totally or partially sheathed by a papery spathe. Flowers from January to April. **Spikelets** up to 4 mm long, occur in pairs, one member pedicellate and awnless, and the other sessile and bearing a twisted awn. **Leaf blade** up to 6 mm wide, flattened, with a distinct midrib and terminating in a sharp point. **Leaf sheath** smooth and round. **Ligule** a short membrane.

Habitat Common in sour grassland and on leached sandy soils in high-rainfall regions. Generally associated with mountains and slopes. In regions with more moderate rainfall, the grass prefers moist soils where water accumulates. **Biome**: Grassland and Savanna.

General An unimportant grass as far as grazing is concerned. Could be regarded as an indicator of acid soils. Palatability decreases as the grass matures. Preferred by oribi. **Grazing value** low. **Ecological status**: mostly Decreaser.

THEMEDA TRIANDRA

Rooigras

A tufted perennial with branched culms, 0,3—1,5 m tall. **Inflorescence** a false panicle consisting of a number of pendulous racemes. Each raceme consists of a group of spikelets supported by a glabrous or hairy spathe. Flowers from October to July. **Spikelets** of two types, the one short and pedicellate with an awn, and the other sessile, up to 7 mm long and awnless. **Leaf blade** up to 8 mm wide, usually flattened, with a distinct midrib. Commonly bluegreen changing from green to reddish brown as it matures. **Leaf sheath** compressed, hairy or glabrous. **Ligule** a membrane, commonly torn in the middle.

Habitat Occurs in all veld types in southern Africa, but is especially common in undisturbed climax grassland. Grows on all soil types. **Biomes**: Grassland, Savanna, Nama-Karoo and Fynbos.

General Although there are differences of opinion, Rooigras is generally regarded as one of the best grazing grasses, especially on the highveld. Leaf production is high and it is very palatable in the young stage. However, its nutritional value is low in winter. Because of its high palatability, its occurence decreases under poor veld management. It is resistant to fire provided that regrowth is not repeatedly overgrazed. Preferred by nearly all grazers. **Grazing value** high to very high. **Ecological status**: Decreaser.

INCONSPICUOUS INFLORESCENCES

* PENNISETUM CLANDESTINUM

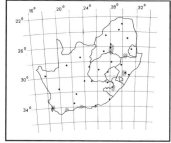

Kikuyu
Kikoejoe

A creeping perennial with stolons and rhizomes, flattened culms up to 300 mm tall and rooting from the lower nodes. **Inflorescence** consists of inconspicuous feathery white structures with 2—4 spikelets enclosed by the upper leaf sheaths. Flowers from August to April. **Spikelets** up to 20 mm long, sessile, white to light green. **Leaf blade** up to 7 mm wide, shortly hairy and initially folded, but later flattened. **Leaf sheath** imbricate. **Ligule** a ring of dense hairs.

Habitat Under natural conditions more widespread in the high-rainfall regions of the country. A grassland species, generally occurring in disturbed areas. Grows on any well drained, fertile soil. **Biomes**: Grassland and Fynbos.

General Originally from East African highlands. A well known lawn grass and an important pasture grass. On fertile soils it provides palatable material which is highly nutritious. Propagated vegetatively or by means of seed. Generally planted as pasture for sheep or dairy cattle. Under optimum conditions it could become a hardy weed which is difficult to control mechanically. Commonly used to stabilize erodible soil on steep slopes along new roads and the embankments of dams. **Grazing value** very high. **Ecological status**: Increaser IIb under natural conditions.

REFERENCES

ACOCKS, J.P.H. 1988. Veld types of South Africa, 3rd edn. **Memoirs of the Botanical Survey of South Africa** No. 57.
BOTHMA, J. du P. (ed.) 1989. **Game ranch management.** J.L. van Schaik, Pretoria.
CHIPPENDALL, L.K.A. 1955. A guide to the identification of grasses in South Africa. **In** Meredith D. (ed.), The grasses and pastures of South Africa, Part 1. Central News Agency, Johannesburg.
CHIPPENDALL, L.K.A. & CROOK, A.O. 1976. **240 grasses of Southern Africa.** M.O. Collins, Salisbury.
DANNHAUSER, C.S. 1980. **'n Sleutel tot die belangrikste veldgrasse van Wes-Transvaal en Noord-Vrystaat.** Dept. of Agriculture and Fisheries, Division Agricultural Information, Pretoria.
DANNHAUSER, C.S. 1987. Aangeplante voergewasse. Special series - **Landbouweekblad**, Johannesburg.
EHLERS, J.H. & COETZEE, J.F. No date. **Description of plant species of economic importance in the R.S.A.** Dept. of Agriculture and Fisheries, Transvaal Region, Pretoria.
FAIR, J. 1986. **John Fair's guide to profitable pastures.** M & J Publications, Harrismith.
GIBBS RUSSELL, G.E., REID, C., VAN ROOY, J. & SMOOK, L. 1985. List of species of Southern African plants, 2nd edn, Part 2. **Memoirs of the Botanical Survey of South Africa** No. 51.
GIBBS RUSSEL, G.E., WATSON, L., KOEKEMOER, M., SMOOK, L., BARKER, N.P., ANDERSON, H.M. & DALLWITZ, M.J. 1990. Grasses of Southern Africa. **Memoirs of the Botanical Survey of South Africa** No. 58.
GRABANDT, K. 1985. **Weeds of crops and gardens in southern Africa.** Ciba-Geigy, Johannesburg.
HäFLIGER, E. & SCHOLZ, H. 1980. **Grass weeds**, Part 1 & 2. Ciba-Geigy, Basel.
HILLIARD, O.M. 1987. **Grasses, sedges, restiads and rushes of the Natal Drakensberg.** University of Natal Press, Pietermaritzburg.
LIGHTFOOT, G. 1975. **Common veld grasses of Rhodesia.** Natural Resources Board of Rhodesia, Salisbury.
LOWREY, T.K. & WRIGHT, S. (eds) 1987. **The flora of the Witwatersrand, Volume 1: The monocotyledonae.** University of the Witwatersrand, Johannesburg.
MüLLER, M.A.N. 1983. **Grasse van Suidwes-Afrika/Namibië.** Directorate Agriculture and Forestry, Windhoek.
ROBERTS, B.R. 1973. **Algemene grasse van die Oranje-Vrystaat.** Provincial Administration of the Orange Free State. Nature Conservation, General Publication No. 3, Bloemfontein.
SMITHERS, R.H.N. 1983. **Mammals of the southern African subregion.** University of Pretoria, Pretoria.
TAINTON, N.M. BRANSBY, D.I. & BOOYSEN, P. DE V. 1976. **Common veld and pasture grasses of Natal.** Shuter & Shooter, Pietermaritzburg.
TAINTON, N.M.(ed.) 1981. **Veld and pasture management in South Africa.** Shuter & Shooter, Pietermaritzburg, in co-operation with University of Natal Press, Pietermaritzburg.
TROLLOPE, W.S.W., POTGIETER, A.L.F. & ZIMBATIS, N. 1989 Assessing veld condition in the Kruger National Park using key grass species. **Koedoe** 32(1): 67—93.
VAN WYK, A.E. & MALAN, S.J. 1988. **Field guide to the wild flowers of the Witwatersrand & Pretoria region.** Struik, Cape Town.

GLOSSARY

Acute - Distinctly and sharply pointed.
Acuminate - Tapering gradually to a long fine point.
Animal unit (AU) - An animal with a mass of 480 kg and which gains 0,5 kg per day on forage with a digestible energy percentage of 55%.
Annual - A plant that completes its life cycle in one year.
Area selective grazing/browsing - Habit of grazing/browsing animals to graze/browse certain areas of the veld/pasture in preference to others.
Arid - A term applied to an area that generally receives little rainfall, hence is a barren or sparsely vegetated area - usually associated with desert conditions.
Articulation - A joint, particularly where a clean break may occur between two parts that drop at maturity, e.g. the lemma and column of some *Aristida* species.
Aspect - Predominant direction of slope of the land.
Auricle - An ear-like outgrowth at the junction of the leaf blade and the leaf sheath.
Awn - A bristle-like appendage particularly of the glume, lemma or palea. See Figure 23.
Axis - A central stem, e.g. of the inflorescence.
Back fire - A fire burning against the wind.
Basal - Situated at the base of an organ or part.
Basal cover - Area of ground covered by the living basal portions of plants.
Biomass - Total amount of living material (plant and animal) present in a particular area at any given time — kg/ha.
Biome - Major regional ecological community of plants and animals.
Bract - A much reduced leaf of the flowering system.
Browsers - Animals that usually feed on leaves, flowers and fruit of woody plant species.
Bulk grazers - Big grazers that normally do not maintain a high degree of selective grazing.

Bushveld - See Savanna.
Camp - Smallest unit to which grazing and/or browsing management is applied.
Canopy cover - Proportion of the ground area covered by the vertical projection of the canopy — %.
Carrying capacity - Potential of an area to support livestock through grazing and/or browsing and/or fodder production over an extended number of years without deterioration of the overall ecosystem — ha/AU or AU/ha.
Climatic climax - Ultimate phase of ecological development of plant communities that the climate of a region will support.
Climax species - A species that is self-perpetuating in the absence of disturbance, with no evidence of replacement by other plant species.
Climax vegetation - Final stable plant community in an ecological succession which is able to reproduce itself indefinitely under existing environmental conditions.
Collar - Junction of the leaf blade and sheath in grasses.
Column - The lower portion of the awn, e.g. the basal part of the awn in some *Aristida* species.
Community - An assemblage of plants growing together and interacting among themselves in a specific location.
Concentrate grazers - In general small grazers that maintain one or other kind of excessive species selective or area selective grazing.
Controlled burning - Burning of veld for a specific reason.
Controlled selective grazing (CSG) - See High production grazing.
Crown fire - A fire that burns in the canopies of trees and shrubs.
Culm - The stem of a grass. See Figure 18.
Cultivated pasture - Pasture that has been established by conventional means involving soil disturbance, removal of

existing vegetation and seedbed preparation.

Decreaser species - Species that dominate in good veld but decrease when veld is mismanaged.

Desiccation - A decrease in available soil moisture due to an increase in runoff. Typically the result of a weakening or degradation in veld condition.

Digestibility - Proportion of a feed that has the potential to be ingested by animals.

Digitate - Arising from one point, like the fingers on a hand, e.g. inflorescence branches in *Dactyloctenium aegyptium*.

Diversity - An expression of the variety of species that exists in a community.

Dominance - Degree of influence that a plant species exerts over a community as measured by its mass or basal area per unit area of the ground surface or by the proportion it forms of the total cover, mass or basal area of the community.

Duplex soil A soil with a relatively permeable topsoil abruptly overlying a very slowly permeable diagnostic horizon.

Ecology - Study of the interrelationships between organisms, and between them and their environment.

Ecosystem - Biological system comprising both living organisms and the abiotic components of the environment.

Ecotype - Plant type or strain within a species, which results from long-term exposure to a particular environment.

Edge effect - Influence of two communities upon their adjoining margins, affecting the composition and density of the population in these bordering areas.

Encroachment - The spread of a plant into an area where previously it did not occur.

Filiform - Cylindrical or thread-like, e.g. leaf blade of *Elionurus muticus*.

Fire intensity - Release of heat energy per unit time per unit length of fire front — kJ/s/m; kW/m.

Fire regime - Season and frequency of burning and the type and intensity of fire.

Flabellate - Fan-shaped; like the arrangement of basal leaf sheaths of some species, e.g. *Eustachys paspaloides*.

Flame height - Perpendicular height of flames from ground level — m.

Floret - A small flower; in grasses consisting of the lodicules, stamens and pistil enclosed by the lemma and palea. See Figure 23.

Frequency of occurrence - Ratio between the number of sample units that contain a plant species and the total number of sample units.

Frequency of burning - Frequency with which fires are applied, expressed as the number of years elapsing between burns, e.g. annual burn, biennial burn.

Fynbos - Veld in which the dominant plants are non-grasses with sclerophyllous, ericoid or cupressoid leaves but in which the associated grasses provide most of the forage material.

Geniculate - With kneelike bends, such as the nodes of some species, e.g. *Eragrostis lehmanniana*.

Glumes - The two lowermost sterile bracts at the base of the spikelet. See Figure 23.

Grain 'seed' - The uncovered and usually hard fruit of the grass plant.

Grassveld - Veld in which grasses are the dominants and provide most of the forage material.

Grazer - An animal that utilizes grazing.

Grazing (n) - That portion of the herbaceous vegetation that is available for consumption by animals.

Grazing (v) - Utilization of herbaceous vegetation by animals.

Habitat - Type of environment in which a plant or animal normally lives.

Head fire - A fire burning with the wind.

Heat yield - Amount of heat energy available for release per unit mass of fuel — kJ/kg.

High performance grazing - See High production grazing.

High production grazing (HPG) - Occupation of a camp by grazing animals until the acceptable grass species have been grazed to a stage that will ensure rapid regrowth and a high production of forage.

High utilization grazing (HUG) - Occupation of a camp by grazing animals until all the grass species have been heavily grazed.

Hydrophyte - Plant usually found growing in water or in soil containing water well in excess of field capacity most of the time.

Increaser I species - Species that dominate in poor veld and increase with understocking of selective grazing.

Increaser Ia species - Species that increase with moderate understocking or selective grazing.

Increaser Ib species - Species that increase with severe understocking or selective grazing.

Increaser II species - Species that dominate in poor veld and increase with overstocking.

Increaser IIa species - Species that increase with light overstocking.

Increaser IIb species - Species that increase with moderate overstocking.

Increaser IIc species - Species that increase with severe overstocking.

Indicator species - Those plant species that best indicate the condition of the veld or pasture, or the nature of the environment.

Inflorescence - The system that bears the spikelets, usually borne at the apex of the culm. See Figure 18.

Internode - The part of the stem between two successive nodes. See Figure 18.

Invader species - Species that are not indigenous to a specific area.

Involute - Having the margins of the leaves rolled inwards, e.g. in *Enneapogon cenchroides*.

Karoo - Veld in which the dominants are xerophytic dwarf shrubs and succulents. Most of the forage material is derived from the dwarf shrubs and the associated xerophytic grasses.

Keel - A ridge like the keel of a boat, e.g. the back of the leaf blade in *Eleusine coracana*.

Lanceolate - Widest in the basal third and gradually tapering to a point. See Figure 20.

Large stock unit - See Animal unit.

Lateral - On, or fixed to, the side of an organ.

Leaf blade - The elongated or expanded limb of the leaf, borne at the apex of the sheath, above the ligule. See Figure 19.

Leaf sheath - The basal part of a leaf, below the ligule, which is usually wrapped around the culm. See Figure 19.

Ligule - A membrane or fringe of hairs situated at the junction of the leaf blade and leaf sheath, facing the culm. See Figure 21.

Linear - Very long and narrow with more or less parallel margins, like the leaf blades of most grasses. See Figure 20.

Lemma - The lower bract enclosing the floret. See Figure 23.

Membranous - Thin and often translucent.

Metabolic mass - Mass of an animal raised to the power of three quarters.

Midrib - The main central vein of a leaf. See Figure 19.

Mixed veld - Veld in which the acceptability of the forage plants is intermediate between that in sweetveld and that in sourveld, thus allowing the veld to be utilized for only a portion of each year.

Node - The usually thickened solid part of the culms or axes from which leaves, bracts or branches arise. See Figure 18.

Non-selective grazing (NSG) - See High utilization grazing.

Nutritive value - Concentration of nutrients in a feed.

Overgrazing - Excessive defoliation of the grass sward by animals to the detriment of the condition of the veld or pasture.

Overstocking - Concentrating livestock in an area to the point where the stocking rate exceeds the carrying capacity of the veld or pasture.

Palatability - Attractiveness of feed to animals as determined by specific factors of the forage.

Palea - The upper bract enclosing the floret. See Figure 23.

Panicle - A branched inflorescence consisting of a primary axis, with secondary axes which bear pedicellate spikelets. See Figure 22.

Pedicel - The stalk of a spikelet. See Figure 22.

Peduncle - The upper part of the culm which carries the inflorescence. See Figure 22.

Perennial - A plant with a life cycle longer than two years.

Period of occupation - Length of time during which a particular camp is being utilized without interruption.

Period of rest - Length of time during the growing season when a camp is afforded uninterrupted rest from grazing and/or browsing for the purpose of restoring the productivity of the veld and/or pasture.

Phytomass - Total mass of plants, including dead attached parts, per unit area — kg/ha.

Pioneer plants - Plants capable of invading bare or disturbed sites and persisting there until replaced by other species.

Plant succession - Progressive development of vegetation in an area through a series of different plant communities, finally terminating in a climax community.

Potential carrying capacity - Potential of a farming unit to support livestock through grazing and/or browsing and/or fodder production when all the factors that affect its productivity are at an optimum level — ha/AU or AU/ha.

Preferred species - Plant species that are preferred and utilized first by animals.

Primary succession - Succession on surfaces exposed for the first time, which have never before borne vegetation.

Raceme - An unbranched inflorescence in which the rachis bears stalked spikelets. See Figure 22.

Rachilla - The axis of a spikelet.

Rachis - The central axis of a raceme or spike. See Figure 22.

Rainfall erosivity - A measure of the erosive force of rainfall. For a particular storm, it is the product value of the kinetic energy and the maximum 30-minute storm intensity.

Rhizome - An underground stem. See Figure 18.

Rotational grazing/browsing - Type of management that requires the grazing/browsing allotted to a group or groups of animals for the entire grazeable/browseable period, to be subdivided into at least one (usually more) camp more than the number of animal groups. It involves successive grazing/browsing of the camps by the animals in a rotation so that at any time the animals are concentrated on as small a part of the grazing/browsing available to them during the entire grazeable/browseable period, as fencing will permit.

Rotational resting - Type of management where the veld and/or pasture is subdivided into at least one more camp than there are groups of animals and involves the successive resting of the camp from grazing and/or browsing for a specific purpose aimed at the restoration of vigour and productivity rather than merely the regrowth of vegetative material for grazing and/or browsing.

Roughage - Plant materials that are relatively high in fibre and low in nutrients.

Rugose - Wrinkled or ridged, like the surface of the lemma in some species, e.g. *Setaria ustilata*.

Savanna - A physiognomic type of vegetation comprising a tree or shrub overstory and an herbaceous understory.

Scabrid, scabrous - Rough to the touch.

Secondary succession - Succession which occurs after the destruction of part or all of the original vegetation on a site.

Sediment delivery ratio - The ratio of sediment leaving the lowest point of a catchment to the average gross soil loss within the same catchment.

Selective grazing/browsing - Selective utilization of the grazing/browse by animals.

Sessile - Without a stalk.

Small stock unit (SSU) - An animal which is equivalent to the 1/6 of a large stock unit, e.g. goat or sheep.

Soil erodibility - The rate of erosion per unit of rainfall erosivity expressed in ton/ha from a unit plot. A unit plot is defined as a plot 22.12 m long on a 9% slope in continuous fallow, tilled up and down the slope.

Soil loss tolerance value (Allowable soil loss) - The maximum long term soil loss rate for a specific soil that will allow sustained economical production in the foreseeable future.

Sourveld - Veld in which the forage plants become unacceptable and less nutritious on reaching maturity, thus allowing the veld to be utilized for only a portion of each year in the absence of licks.

Spathe - A bract which encloses or supports the inflorescence or a part of it.

Species abundance - Total number of individuals of a species in an area, population or community.

Species composition - Relative proportion of different plant species occurring in a specific area.

Species selective grazing/browsing - Habit of grazing/browsing animals to graze/browse certain species of the vegetation in preference to others.

Spike - An inflorescence in which the rachis carries sessile spikelets. See Figure 22.

Spikelet - The basic flower bearing unit of the grass inflorescence consisting of two glumes and one or more florets. See Figure 23.

Stability - The tendency of an ecological system to resist change when subjected to fluctuating conditions.

Standing crop - Total amount of aboveground plant material per unit area.

Stocking density - Concentration of livestock on the veld and/or pasture at any instant in time — AU/ha.

Stocking rate - Area of land in the system of management which the operator has allotted to each animal unit in the system, and is expressed per length of the grazeable and/or browseable period of the year — ha/AU or AU/ha.

Stolon - A stem that creeps above the ground, rooting at the nodes and giving rise to new culms. See Figure 18.

Surface fire - A fire that burns in the surface fuels.

Surface fuels - All combustible material on the soil surface occurring as standing grass, shrublets, seedlings, forbs, fallen leaves, twigs and bark.

Sweetveld - Veld in which the forage plants retain their acceptability and nutritive value after maturity or in which different plants are acceptable at different times so that the veld can be utilized by stock at all times of the year.

Terminal - At the tip or end.

Trampling - Effect of hoof action by ungulates on herbaceous plants and the soil surface.

Undergrazing - Underutilization of the herbaceous component of the vegetation.

Understocking - Stocking on area with animals so that the stocking rate is less than the carrying capacity of the veld or pasture.

Unilateral - The arrangement of spikelets on only one side of the axis, e.g. inflorescence in *Harpochloa falx*.

Veld - Indigenous vegetation used as grazing and/or browsing which may be composed of any of a number of plant growth forms and not necessarily climax vegetation.

Veld condition - Condition of the vegetation in relation to some functional characteristic, normally maximum forage production and resistance to soil erosion.

Veld management - Management of natural vegetation for specific objectives related to different forms of land use.

Veld management practices - All treatments applied in the management of the veld.

Veld management system - Formalized programme of veld management through which veld management practices are applied.

Veld reclamation - See Veld rehabilitation.

Veld rehabilitation - Restoration of degraded veld to a productive and stable condition.

Veld type - Unit of vegetation whose range of variation is small enough to permit the whole of it to have the same farming potentialities.

Whorl - A ring of similar parts radiating from a single node.

INDEX TO PLANT NAMES

The plant name index is divided into three parts to make quick and easy reference possible i.e. the index to botanical names, Afrikaans common names and English common names.

The index to botanical names comprises three types of entries:
Main index entries: This entry refers to all the species which are described and illustrated. These names are shown in **bold type** and the page number in **bold** refers to the specific page where the species is described. Exotic species are indicated with an asterisk.
Synonyms: All the species which have had changes to their names during the past 10 years are included to assure a certain degree of continuity. These entries are in *italic type* and the page number refers to the species description. The synonym is however not repeated on this page.
Incidental entries: In some cases reference had to be made to a species not described in this book. These reference species are shown in normal type.
  Example of index entries in the botanical name index:
  Panicum maximum   236, 237, **239**   - A main entry with the description of the species on page 239 and reference to the species on pages 236 and 237.
  * **Pennisetum clandestinum**   280   - The asterisk indicates an exotic species.
  Andropogon amplectens   129   - A synonym described on page 129 under its current name.
  Eragrostis tenella   109   - A species not described in this book, referred to on page 109.

The **Afrikaans and English name indexes**, comprise two types of entries:
  The most common name is shown in **bold** type. Other common names or synonyms are shown in *italic type*, and the page number refers to the species description. Only the most common name is however repeated with the description.
  Example of index entries in the common name indexes:
  L.M. Grass   126   - the most common name with illustration and description on page 126.
  Natal Crowfoot   126 - another common name or synonym with description and illustration on page 126.

INDEX TO BOTANICAL NAMES

A

Acoceras macrum 161
Agrostis barbuligera 210
Agrostis eriantha var. **eriantha** 210
* **Agrostis montevidensis** 211
Alloteropsis semialata subsp. **eckloniana** 133, 208
Alloteropsis semialata subsp. semialata 133
Andropogon amplectens 129
Andropogon appendiculatus 134
Andropogon chinensis 124, 125
Andropogon eucomus 264, 266
Andropogon filifolius 130
Andropogon gayanus var. **polycladus** 265
Andropogon huillensis 134, 264, **266**
Andropogon schinzii 124
Andropogon schirensis 124, **125**, 130, 158
Anthephora angustifolia 74
Anthephora argentea 74, 77
Anthephora pubescens 74, **75**, 77
Apochaete hispida 208
Aristida adscensionis 102
Aristida alopecuroides 103
Aristida andoniensis 180
Aristida barbicollis 213
Aristida bipartita 212, 215
Aristida canescens subsp. **canescens** 179
Aristida capensis var. *dieterleniana* 206
Aristida ciliata 204
Aristida congesta subsp. **barbicollis** 213
Aristida congesta subsp. **congesta** 103
Aristida curvata 102
Aristida diffusa subsp. **burkei** 214
Aristida diffusa subsp. diffusa 214
Aristida effusa 212
Aristida graciliflora 105
Aristida junciformis subsp. **junciformis** **104**, 179
Aristida longicauda 103
Aristida meridionalis 254
Aristida obtusa 205
Aristida pilgeri 255
Aristida rhiniochloa 180
Aristida scabrivalvis subsp. contracta 215
Aristida scabrivalvis subsp. **scabrivalvis** 212, **215**
Aristida sciurus 255
Aristida sericans 206

Aristida spectabilis 255
Aristida stipitata subsp. **graciliflora** **105**, 106
Aristida stipitata subsp. **stipitata** 105, **106**
Aristida submucronata 102
Aristida uniplumis var. *uniplumis* 251
Aristida vestita 214, 254
Arundinella nepalensis 107
Asthenatherum glaucum 108
Avena barbata 216
* **Avena fatua** 216
Avena sterilis 216

B

Bewsia biflora 162
Bothriochloa bladhii 154, **217**
Bothriochloa glabra 217
Bothriochloa insculpta 153, **154**, 217
Bothriochloa insculpta var. *vegetior* 217
Bothriochloa pertusa 153
Bothriochloa radicans 153, **154**
Brachiaria arrecta 167
Brachiaria brizantha 163
Brachiaria deflexa 218
Brachiaria eruciformis 164
Brachiaria marlothii 165
Brachiaria nigropedata 166
Brachiaria serrata 167
Brachiaria xantholeuca 167
* **Briza maxima** 219, **256**
* **Briza minor** 219
* **Bromus catharticus** 257
* **Bromus diandrus** 258
Bromus unioloides 257
Bromus willdenowii 257

C

Cenchrus ciliaris 76, 186
Centropodia glauca 108
* **Chloris gayana** 135, **136**, 142, 152
Chloris mossambicensis 136
Chloris pycnothrix 136, **137**
Chloris virgata 136, **138**
Chrysopogon montanus var. *tremulus* 181
Chrysopogon serrulatus 181
Coelachyrum yemenicum 168, 169

Ctenium concinnum   117
Cymbopogon afronardus   269
Cymbopogon excavatus   129, **267**, 268
Cymbopogon plurinodis   268
Cymbopogon validus   269
Cymbosetaria sagittifolia   244
Cynodon aethiopicus   140
Cynodon dactylon   139
* **Cynodon nlemfuensis**   140
Cypholepis yemenica   168

D

Dactyloctenium aegyptium   141, 157
Dactyloctenium australe   126, 127
Dactyloctenium geminatum   126, **127**
Dactyloctenium giganteum   142
Danthonia disticha   83
Danthonia glauca   108
Dichanthium annulatum var. **papillosum**   153, **155**, 156
* **Dichanthium aristatum**   153, **156**
Dichanthium nodosum   155
Dichanthium papillosum   155
Digitaria argyrograpta   143
Digitaria brazzae   **144**, 149
Digitaria ciliaris   147
Digitaria diagonalis var. **diagonalis**   128, 146, **182**
Digitaria eriantha   145
Digitaria eylesii   **128**, 146, 182
Digitaria maitlandii   146, **182**
Digitaria monodactyla   77, **118**
Digitaria pentzii   145
* **Digitaria sanguinalis**   147
Digitaria smutsii   145
Digitaria ternata   146, **148**
Digitaria tricholaenoides   144, **149**
Digitaria trichopodia   182
Digitaria uniglumis   182
Digitaria velutina   150
Diheteropogon amplectens   124, 125, **129**, 130
Diheteropogon filifolius   130
Dinebra retroflexa var. **condensata**   183
Diplachne biflora   162
Diplachne eleusine   168, **169**
Diplachne fusca   184
Dolichochaete rehmannii   209

E

Echinochloa colona   170
Echinochloa crus-galli   170
Ehrharta calycina   220
Ehrharta villosa var. **maxima**   112, **185**
Ehrharta villosa var. villosa   185
Eleusine africana   157
Eleusine coracana subsp. **africana** 141, **157**
Eleusine indica subsp. *africana   157*
Elionurus argenteus   77
Elionurus muticus   74, **77**, 88, 118
Enneapogon cenchroides   76, **186**, 187, 201
Enneapogon filifolius   187
Enneapogon scoparius   76, **187**
Enteropogon macrostachyus   119
Enteropogon monostachyus subsp. **africanus**   119
Enteropogon rupestris   119
Eragrostis abyssinica   259
Eragrostis arenicola   109
Eragrostis aspera   221
Eragrostis atherstonei   232
Eragrostis biflora   222
Eragrostis brizantha   171
Eragrostis capensis   188, **193, 194**
Eragrostis chalcantha   193
Eragrostis chloromelas   223, 224, 227
Eragrostis cilianensis   229
Eragrostis ciliaris   109
Eragrostis curvula   223, **224**
Eragrostis cylindriflora   232
Eragrostis denudata   190
Eragrostis echinochloidea   171, 228
Eragrostis gummiflua   189, 233
Eragrostis henrardii   232
Eragrostis heteromera   225
Eragrostis lehmanniana var. **lehmanniana**   226, 230, 232
Eragrostis micrantha   227
Eragrostis nindensis   190, 193
Eragrostis obtusa   171, **228**
Eragrostis pallens   191
Eragrostis plana   192, 203, 231
Eragrostis pseudoscleranta   229
Eragrostis racemosa   188, 190, **193**, 229
Eragrostis rigidior   230
Eragrostis robusta   224
Eragrostis rotifer   225
Eragrostis superba   188, **194**
* **Eragrostis tef**   259
Eragrostis tenella   109
Eragrostis tenuifolia   229, **231**
Eragrostis trichophora   226, **232**
Eragrostis viscosa   109, 189, **233**
Eragrostis volkensii   263
Eriochloa borumensis   195
Eriochloa meyeriana subsp. meyeriana   195
Eriochloa stapfiana   195
Eulalia villosa   151, 159
Eustachys mutica   152
Eustachys paspaloides   135, 142, **152**

F

Festuca caprina 234
Festuca costata 234
Festuca scabra 234
Fingerhuthia africana 78
Fingerhuthia sesleriiformis 78

H

Harpochloa falx 120
Helictotrichon dodii 110
Helictotrichon longifolium 110
Helictotrichon natalense 110
Helictotrichon turgidulum 110
Heteropogon contortus 79, 89, 98, 101
* **Hordeum murinum** subsp. **murinum** 80
Hyparrhenia anamesa 270
Hyparrhenia collina 275
Hyparrhenia cymbaria 271
Hyparrhenia dichroa 272
Hyparrhenia dissoluta 276
Hyparrhenia dregeana 275
Hyparrhenia filipendula 270, 274
Hyparrhenia filipendula var. filipendula 273
Hyparrhenia filipendula var. **pilosa 273**
Hyparrhenia glauca 275
Hyparrhenia hirta 270, 273, **274**
Hyparrhenia rufa 272
Hyparrhenia tamba 275
Hyparrhenia variabilis 271
Hyperthelia dissoluta 276

I

Imperata cylindrica 81
Ischaemum afrum 90, **158**
Ischaemum arcuatum 159
Ischaemum brachyatherum 158
Ischaemum fasciculatum 151, **159**
Ischaemum glaucostachyum 158

L

* **Lagurus ovatus 111**
Lolium multiflorum 82
* **Lolium perenne 82**
Loudetia flavida 196
Loudetia simplex 196

M

Melica brevifolia 113
Melica decumbens 112, 113
Melica neesii 112

Melica pumila 113
Melica racemosa 112, **113**
Melinis nerviglumis 197, 235
Melinis repens subsp. **repens** 197, **235**
Merxmuellera disticha 83
Microchloa caffra 118, **121**
Microchloa kunthii 121
Monocymbium ceresiiforme 277

O

Oplismenus hirtellus 172
Oplismenus undulatifolius 172

P

Panicum coloratum var. **coloratum 236**, 239
Panicum deustum 237, 239
Panicum dregeanum 238
Panicum fulgens 240
Panicum laevifolium var. *laevifolium 241*
Panicum maximum 236, 237, **239**
Panicum natalense 240
Panicum schinzii 241
Panicum stapfianum 236
Panicum volutans 242
Paspalum commersonii 174
* **Paspalum dilatatum 173**, 260
Paspalum distichum 131
* **Paspalum notatum** 131, **132**, 174
Paspalum orbiculare 174
Paspalum paspalodes 131
Paspalum scrobiculatum 174
* **Paspalum urvillei** 173, **260**
Paspalum vaginatum 131
* **Pennisetum clandestinum 280**
Pennisetum macrourum 85
* **Pennisetum setaceum** 84, 86
Pennisetum sphacelatum 78, **85**
* **Pennisetum villosum** 84, **86**
Pentaschistis curvifolia 198
Pentaschistis pallida 198
Perotis patens 87
* **Poa annua 243**
Pogonarthria squarrosa 199
Pseudobrachiaria deflexa 218

R

Rhynchelytrum nerviglume 197
Rhynchelytrum repens 235
Rhynchelytrum setifolium 197

S

Schizachyrium jeffreysii   77, **88**
Schizachyrium sanguineum   **89**
Schizachyrium semiberbe   *89*
Schmidtia bulbosa   *201*
Schmidtia kalihariensis   186, **200**
Schmidtia pappophoroides   200, **201**
Sehima galpinii   **90**
Setaria anceps   *94*
Setaria chevalieri   *261*
Setaria flabellata subsp. *natalensis*   *96*
Setaria homonyma   **202**
Setaria incrassata   91, 92
Setaria insignis   *261*
Setaria lindenbergiana   **114**, 261
Setaria megaphylla   114, 261
Setaria neglecta   *95*
Setaria nigrirostris   91, **92**
Setaria pallide-fusca   93, 94, 97
Setaria palustris   *91*
Setaria perennis   *95*
Setaria plicatilis   114
Setaria sagittifolia   **244**
Setaria sphacelata var. **sericea**   94, 95
Setaria sphacelata var. **sphacelata**   **95**
Setaria sphacelata var. **torta**   **96**
Setaria ustilata   93, **97**
Setaria verticillata   **115**
Setaria woodii   *91*
Sorghum bicolor subsp. **arundinaceum**   **245**
Sorghum bicolor subsp. drummondii   245
Sorghum halepense   245
Sorghum versicolor   **262**
Sorghum verticilliflorum   *245*
Sporobolus africanus   **116**, 192, 203, 246
Sporobolus capensis   *116*
Sporobolus centrifugus   249
Sporobolus coromandelianus   248
Sporobolus festivus   250
Sporobolus fimbriatus   116, 203, **246**
Sporobolus ioclados   **247**
Sporobolus ludwigii   248
Sporobolus marginatus   *247*
Sporobolus natalensis   116, 246
Sporobolus nitens   **248**
Sporobolus pectinatus   **249**
Sporobolus pyramidalis   116, 192, 199, **203**, 246
Sporobolus smutsii   *247*
Sporobolus stapfianus   **250**
Sporobolus usitatus   *247*
Stenotaphrum secundatum   **122**
Stipagrostis ciliata var. **capensis**   **204**, 205
Stipagrostis obtusa   204, **205**
Stipagrostis uniplumis var. neesii   251
Stipagrostis uniplumis var. **uniplumis**   **251**
Stipagrostis zeyheri subsp. **sericans**   **206**

T

Tetrachne dregei   **175**
Tetrapogon mossambicensis   *136*
Themeda triandra   89, 208, 209, **278**
Trachypogon capensis   *98*
Trachypogon spicatus   79, 90, **98**, 101
Tragus berteronianus   **99**, 100
Tragus racemosus   99, **100**
Tricholaena monachne   **252**
Trichoneura grandiglumis   **253**
Trichopteryx dregeana   **263**
Triraphis andropogonoides   **207**
Triraphis schinzii   207
Tristachya hispida   *208*
Tristachya leucothrix   **208**
Tristachya rehmannii   **209**

U

Urelytrum agropyroides   90, 98, **101**
Urelytrum squarrosum   *101*
Urochloa bolbodes   *160*
Urochloa mosambicensis   160, **176**
Urochloa oligotricha   **160**, 176
Urochloa panicoides   **177**
Urochloa pullulans   *176*
Urochloa rhodesiensis   *176*
Urochloa ruschii   **177**
Urochloa stolonifera   160, 176

INDEX TO AFRIKAANS COMMON NAMES

A

Afrikaanse osgras 157
Agrostis
   Grootpluim-agrostis **210**
Agtdaepluimgras 190
Andropogon
   Breëblaar-andropogon 129
   Grootwitbaard-andropogon 266
   Kleinwitbaard-andropogon 264
Assegaaigras **79**
Augustinus-gras 122

B

Baangras 158
Bahiagras **132**
Bastersinjaalgras **218**
Beesgras 107
   Bosveldbeesgras **176**
   Meerjarige Beesgras **160**
   Tuinbeesgras **177**
Bergblinkgras 197
Berggras 275
Bergkoperdraad **83**
Bergsoetgras 224
Besemdrieblomgras **209**
Besemgras **191**, 196, 207, 209
Blinkaarboesmangras **251**
Bloubuffelsgras **76**
Blougras 134, **265**
   Breëblaarblougras **129**
   Eenjarige Blougras **243**
   Harige-blougras **124**
   Rhodesiese-blougras 265
   Smalblaarblougras **130**
   Vleiblougras **134**
Blousaad-tricholaena 252
Blousaadgras 162, 232, 246, **252**
Blousaadsoetgras 239
Bloutamboekiegras **275**
Boesmangras
   Blinkaarboesmangras **251**
   Drieveerboesmangras **206**
   Kortbeenboesmangras **205**
   Langbeenboesmangras **204**
Bokbaardgras **98**
Bootjiegras **277**

Bootjietamboekiegras **271**
Borseltjiegras **75**
   Silwerborseltjiegras **74**
Bosgras **172**
Bosluisgras **171**
Bosveldbeesgras **176**
Bosveldfynsaadgras **246**
Brakvleigras 184
Breëblaarblougras **129**
Breëblaarbuffelsgras **237**
Breëblaarterpentyngras **267**
Breëkrulblaar **230**
Bronchogras 257
Broodgras 163
Broodsinjaalgras **163**
Bruin hoenderspoor **152**
Bruinkruisgras 152
Bruinsaadvingergras **182**
Bruinvingergras **144**
Buffelsgras 239
   Bloubuffelsgras **76**
   Bosbuffelsgras 114
   Breëblaarbuffelsgras **237**
   Gewone Buffelsgras **239**
   Katstertbuffelsgras 76
   Kleinbuffelsgras 236
   Landbuffelsgras 241
   Natal-buffelsgras 240
   Persbuffelsgras **238**
   Rietbuffelsgras 237
   Soetbuffelsgras **241**
   Strandbuffelsgras **122**
   Suurbuffelsgras **240**
   Vleibuffelsgras 241
   Witbuffelsgras **236**
Bulgras **85**

D

Dekgras **274**
   Bloudekgras 275
   Harige-rooidekgras 88
   Rooidekgras 89
   Turfdekgras 90
Donkersaadgras **133**
Donsgras **81**
Doppiesgras **185**
Douvatgras **228**

Draadbloustam 130
Drakensberg-vingergras 146
Drieblomgras
    Besemdrieblomgras 209
    Harige-drieblomgras 208
Drieveerboesmangras 206
Dronkgras 112

E

Eenjarige Steekgras 102
Eenjarige Blougras 243
Eenvingergras 118
Ehrharta
    Gewone Ehrharta 220
Elastiese-eragrostis 231
Elsgras 121
Eragrostis
    Elastiese-eragrostis 231
    Grootpluim-eragrostis 221
    Hartjie-eragrostis 188
    Kleef-eragrostis 189
    Klewerige-eragrostis 233
    Krulblaar-eragrostis 230
    Platsaad-eragrostis 194
    Skadu-eragrostis 222
    Taaipol-eragrostis 192
    Vals-eragrostis 162
    Voetpad-eragrostis 229
    Wollerige-eragrostis 109

F

Finessegras 227
Flossiegras 113
Fluweelgras 167, 235
    Geelfluweelgras 151
Fluweelsinjaalgras 167
Fynblousaadgras 252
Fynsaadgras
    Bloufynsaadgras 250
    Bosveldfynsaadgras 246
    Gewone Fynsaadgras 246
    Katstert-fynsaadgras 203
    Krulblaarfynsaadgras 248
    Panfynsaadgras 247
    Rotstert-fynsaadgras 116
    Smalpluimfynsaadgras 203
    Veselfynsaadgras 250
Fyntamboekiegras 273

G

Geelfluweelgras 151
Geeltamboekiegras 276
Geelturfgras 90

Gerftamboekiegras 270
Gewone Buffelsgras 239
Gewone Ehrharta 220
Gewone Mannagras 95
Gewone Paspalum 173
Gewone Urochloa 176
Gewone Wildehawer 216
Gewone Wildesorghum 245
Gewone Wortelsaadgras 99
Ghagras 108
Gomgras 189
Goue-mannagras 94
Gouebaardgras 181
Grootbewertjiegras 256
Grootklokkiegras 256
Grootpluim-agrostis 210
Grootpluim-eragrostis 221
Grootrolgras 212
Grootwitbaardgras 266
Grootwortelsaadgras 100

H

Haarwurmgras 86
Haasstert 111
Hamelgras 190
Harige-blougras 124
Harige-drieblomgras 208
Harige-pluimgras 232
Harige-herfsgras 88
Harpoengras 195
Hartjiesgras 188, *193*
Hawergras
    Kleinhawergras 110
Herfsgras
    Harige-herfsgras 88
    Rooiherfsgras 89
Hoenderspoor 141
    Bruin Hoenderspoor 152
    Gewone Hoenderspoor 141
    Reuse Hoenderspoor 142

I

Indiese osgras 157

K

Kalahari-suurgras 200
Kalaharisandkweek 201
Kalkgras 187
Kammetjiesgras 249
Katstertfynsaadgras 203
Katstertgras 87, *183*
Katstertsteekgras 103

Katstertturfgras 183
Kikoejoe 280
Kinagras 101
Klein rolgras 253
Kleinbewertjiegras 219
Kleinhawergras 110
Kleinkapokgras 264
Kleinklokkiegras 219
Kleinwitbaardgras 264
Klewerige-eragrostis 233
Klitsgras *100,* 115
   Kousklits 99
Klokkiegras
   Grootklokkiegras 256
   Kleinklokkiegras 219
Klosgras 153
Klossiegras *138*
   Blouklossiegras 217
   Persklossiegras 217
Knietjiesgras 226
Koperdraadgras 214
Koperdraad 77
   Bergkoperdraad 83
Kortbeenboesmangras 205
Kortbeenskubgras 168
Kropaargras 175
Kruipgras 165
Kruipmannagras 96
Kruisgras 182
Kruisvingergras 147
Krulblaar 223
   Breëkrulblaar 230
Krulblaarfynsaadgras 248
Krulgras 181
Kuilgras 184
Kwasgras 198
Kweek 139
   Bankrotkweek 131
   Sandkweek 201
   Reuse Kweekgras 140
Kweekpaspalum 131

L

L.M.-gras 126
Langbeen-paspalum 260
Langbeenboesmangras 204
Langbeenskubgras 169
Langbeensteekgras 254
Langnaald-bromus 258
Langnaaldsteekgras 105, 106
Langsteekgras 255
Lemoengras 267
Litjiesgras 161
Litjiesinjaalgras 164
Lossteekgras 213

M

Mannagras
   Gewone Mannagras 95
   Gouemannagras 94
   Katstertmannagras 95
   Kleinkruipmannagras 96
   Kruipmannagras 96
   Riffelblaarmannagras 261
   Skadumannagras 97
   Swartsaadmannagras 92
   Tuinmannagras 93
   Vleimannagras 91
   Waaierblaarmannagras 202
Matgras 263
Meerjarige Beesgras 160
Meerjarige Raaigras 82
Meerjarige Spinnerakgras 136
Misbeltgras 211
Moerasgras 170
Munnik-swenkgras 234

N

Naaldgras 119
Natal-rooipluim 235
Negenaaldgras 186
Ngongoni-steekgras 104
Nylgras 161

O

Osgras 157
Oulandsgras 224

P

Panfynsaadgras 247
Paspalum
   Buffelskweek-paspalum 131
   Commerson-paspalum 174
   Gewone Paspalum 173
   Kweekpaspalum 131
   Langbeen-paspalum 260
   Reuse-paspalum 260
   Vasey-paspalum 260
   Veldpaspalum 174
Perdegras 207
Persaar 87
Persbuffelsgras 238
Persklossiegras 217
Perssteekgras 215
Persvingergras 149
Pluimsekelgras 199
Predikantsluis 258

Pronkgras 84
Pylblaargras 244
Pylgras 79
   Reuse-pylgras 98

R

Randjiesgras 114
Reënboogvleigras 156
Reddingsgras 257
Reuse Hoenderspoor 142
Reuse Terpentyngras 269
Rhodesgras 135
Rietgras 107, 266
Riffelblaarmannagras 261
Riviergras 107
Rolgras 242
   Grootrolgras 212
   Klein rolgras 253
Rooigras 278
Rooiherfsgras 89
Rooikopergras 225
Rooipluim
   Natal-rooipluim 235
Rooisaadgras 208
Rooitamboekiegras 272
Rooivleigras 159
Ruspergras 120

S

Sandkweek 201
Sekelgras 117, 199
Setaria
   Berg-setaria 114
   Bleëblaar-setaria 261
   Goue-setaria 94
   Klits-setaria 115
   Kruip-setaria 96
   Riffelblaar-setaria 261
   Swartsaad-setaria 92
   Tuin-setaria 93
   Vlei-setaria 91
   Waaierblaar-setaria 202
Silweraargras 81
Silwerborseltjiegras 74
Silwervingergras 143
Sinjaalgras
   Bastersinjaalgras 218
   Broodsinjaalgras 163
   Fluweelsinjaalgras 167
   Litjiesinjaalgras 164
Skadu-eragrostis 222
Skadumannagras 97
Skurwesteekgras 180
Slapvingergras 150

Smalblaarblougras 130
Smalblaarterpentyngras 268
Smalhartjiesgras 193
Smutsvingergras 145
Soetbuffelsgras 241
Spinnerakgras 137
   Meerjarige Spinnerakgras 136
Steekblaarblinkgras 197
Steekgras
   Aapstertsteekgras 103
   Assegaaisteekgras 104
   Eenjarige Steekgras 102
   Katstertsteekgras 103
   Koperdraadsteekgras 254
   Langbeensteekgras 254
   Langnaaldsteekgras 105, 106
   Langsteekgras 255
   Lossteekgras 213
   Ngongoni-steekgras 104
   Perssteekgras 215
   Skurwesteekgras 180
   Vaalsteekgras 179
   Witsteekgras 213
Stergras 140
Sterretjiesgras 142
Stingelgras 196
Stinkgras 154
Stippelgras 153
Strandbuffelsgras 122
Suurbuffelsgras 240
Suurgras 200
   Kalahari-suurgras 200
   Vaalsuurgras 186
Suurpol 77
Swartsaadgras 133, 262
Swartsaadmannagras 92
Swartsaadtweevingergras 128
Swartsaadvingergras 148
Swartvoetjiegras 166
Swartwildesorghum 262
Sygras 81

T

Taaipol 116, *192, 203*
Taaipol-eragrostis 192
Tamboekiegras
   Bloutamboekiegras 275
   Bootjietamboekiegras 271
   Dektamboekiegras 274
   Fyntamboekiegras 273
   Geeltamboekiegras 276
   Gerftamboekiegras 270
   Rooitamboekiegras 272
Tef 259

Terpentyngras
    Breëblaarterpentyngras 267
    Gewone Terpentyngras 267
    Reuse Terpentyngras 269
    Smalblaarterpentyngras 268
Tuin-urochloa 177
Tuinbeesgras 177
Tuinmannagras 93
Turfgras 158
    Geelturfgras 90
    Katstertturfgras 183
Tweevingergras 125

V

Vaalgras 201
Vaalsteekgras 179
Vals-eragrostis 162
Varkstertgras 101
Veergras 264
Veldpaspalum 174
Veselfynsaadgras 250
Vingergras 145
    Bruinsaadvingergras 182
    Bruinvingergras 144
    Drakensberg-vingergras 146
    Eenvingergras 118
    Kleinvingergras 143
    Kruisvingergras 147
    Persvingergras 149
    Silwervingergras 143
    Slapvingergras 150
    Smutsvingergras 145
    Swartsaadtweevingergras 128
    Swartsaadvingergras 148
    Tweevingergras *124,* 125 *158*
    Vleivingergras 155
    Wolvingergras 145
Vingerhoedgras 78

Vleiblougras 134
Vleigras *203,* 263
    Reënboogvleigras 156
    Rooivleigras 159
Vleimannagras 91
Vleivingergras 155
Voetpad-eragrostis 229

W

Waaierblaarmannagras 202
Waaigras 253
Watergras 173
Watergras 170
Weeluisgras 194
Wildegars 80
Wildehawer
    Gewone Wildehawer 216
    Wildehawergras 277
Wildesorghum
    Gewone Wildesorghum 245
    Swartwildesorghum 262
Wintergras 243
Witbaardgras
    Grootwitbaardgras 266
    Kleinwitbaardgras 264
Witbuffelsgras 236
Witpluim 138
Witpluim-chloris 138
Wollerige-eragrostis 109
Wortelsaadgras
    Gewone Wortelsaadgras 99
    Grootwortelsaadgras 100
Wysergras 127

Y

Ystergras 214

INDEX TO ENGLISH COMMON NAMES

A

African Goose Grass   157
Agrostis
   Large Panicle Agrostis   210
Andropogon
   Broad-leaved Andropogon   129
   Large Silver Andropogon   266
   Small Silver Andropogon   264
   Thread-leaved Andropogon   130
Annual Blue Grass   243
Annual Three-awn   102
Arrow Grass   244
Autumn Grass
   Red Autumn Grass   89
   Silky Autumn Grass   88

B

Bahia Grass   132
Basket Grass   172
Beard Grass   266
Big Quaking Grass   256
Black-seed Bristle Grass   92
Black Sudan Grass   262
Black-seed Wild Sorghum   262
Black-seed Grass   133
Black-footed Brachiaria   166
Black-footed Signal Grass   166
Black-seed Finger Grass   148
Blue Buffalo Grass   76
Blue Grama   155
Blue Grass   265
   Annual Blue Grass   243
   Hairy Blue Grass   124
   Rhodesian Blue Grass   265
Blue-seed Grass   252
Blue-seed Tricholaena   252
Blue Thatching Grass   275
Bluestem   134
   Broad-leaved Bluestem   129
   Thread-leaved Bluestem   130
   Vlei Bluestem   134
   Wire Bluestem   130
Boat Grass   277
Boat Thatching Grass   271
Bottlebrush Grass   187

Bristle Grass
   Annual Bristle Grass   102
   Black-seed Bristle Grass   92
   Brack Swamp Grass   184
   Broad-leaved Bristle Grass   261
   Bur Bristle Grass   115
   Common Bristle Grass   95
   Creeping Bristle Grass   96
   False Bristle Grass   85
   Fan-leaved Bristle Grass   202
   Garden Bristle Grass   93
   Golden Bristle Grass   94
   Mountain Bristle Grass   114
   Ngongoni Bristle Grass   104
   Ribbon Bristle Grass   261
   Shade Bristle Grass   97
   Sticky Bristle Grass   115
   Tassel Bristle Grass   103
   Vlei Bristle Grass   91
Bristle-leaved Red Top   197
Broad-leaved Bluestem   129
Broad-leaved Bristle Grass   261
Broad-leaved Curly Leaf   230
Broad-leaved Panicum   237
Broad-leaved Turpentine Grass   267
Bronze Love Grass   225
Broom Grass   191, 209
Broom Love Grass   191
Broom Trident Grass   209
Brown Finger Grass   144
Brown-seed Finger Grass   182
Buffalo Grass   239
   Blue Buffalo Grass   76
   Coastal Buffalo Grass   122
   Natal Buffalo Grass   240
   Sweet Buffalo Grass   241
   Vlei Buffalo Grass   241
   White Buffalo Grass   236
Bundle Thatching Grass   270
Bur Bristle Grass   115
Bushman Grass
   Cape Bushman Grass   206
   Silky Bushman Grass   251
   Small Bushman Grass   205
   Tall Bushman Grass   204
Bushveld Dropseed   246
Bushveld Signal Grass   176

C

Cape Bushman Grass   206
Carpet Grass   263
Carrot-seed Grass
    Common Carrot-seed Grass   99
    Large Carrot-seed Grass   100
    Small Carrot-seed Grass   99
Cat's Tail   87
Caterpillar Grass   120
Catstail Dropseed   203
Catstail Grass   183
Catstail Vlei Grass   183
Centipede Grass   101
Coastal Buffalo Grass   122
Cocksfoot   175
Commerson Grass   174
Common Bristle Grass   95
Common Carrot-seed Grass   99
Common Crowfoot   141
Common Ehrharta   220
Common Paspalum   173
Common Russet Grass   196
Common Signal Grass   163
Common Thatching Grass   274
Common Wild Oat   216
Common Wild Sorghum   245
Cottonwool Grass   81
Couch Grass   139
Couch Paspalum   131
Crab Finger Grass   147
Creeping Bristle Grass   96
Creeping Grass   165
Crowfoot
    Common Crowfoot   141
    Giant Crowfoot   142
Curly Leaf   223
    Broad-leaved Curly Leaf   230
Curly-leaved Dropseed   248

D

Dallis Grass   173
Dew Grass   228
Drakensberg Finger Grass   146
Dropseed
    Bushveld Dropseed   246
    Catstail Dropseed   203
    Common Dropseed   246
    Fibrous Dropseed   250
    Fringed Dropseed   249
    Narrow-plumed Dropseed   203
    Pan Dropseed   247
    Ratstail Dropseed   116
Duck Grass   141
Dune Ehrharta   185

Durban Grass   126

E

Ehrharta
    Common Ehrharta   220
    Dune Ehrharta   185
Elastic Love Grass   231
Eyles' Finger Grass   128

F

False Barley   80
False Bristle Grass   85
False Love Grass   162
False Signal Grass   218
Fan Grass   152
Fan-leaved Bristle Grass   202
Feather Top   86
Feather-top Chloris   138
Feathered Chloris   138
Fibrous Dropseed   250
Fine Thatching Grass   273
Finesse Grass   227
Finger Grass   145
    Black-seed Finger Grass   148
    Brown Finger Grass   144
    Brown-seed Finger Grass   182
    Common Finger Grass   145
    Crab Finger Grass   147
    Drakensberg Finger Grass   146
    Eyles' Finger Grass   128
    Flaccid Finger Grass   150
    Long-plumed Finger Grass   150
    One-finger Grass   118
    Purple Finger Grass   149
    Silver Finger Grass   143
    Small Finger Grass   143
    Smuts' Finger Grass   145
    Vlei Finger Grass   155
Flaccid Finger Grass   150
Fluffy Grass   113
Fog Grass   211
Footpath Love Grass   229
Forest Grass   172
Fountain Grass   84
Foxtail Grass   76
Fringed Dropseed   249

G

Garden Bristle Grass   93
Garden Signal Grass   177
Garden Urochloa   177
Gha Grass   108
Giant Crowfoot   142

Giant Paspalum   260
Giant Spear Grass   98
Giant Three-awn   254
Giant Turpentine Grass   269
Golden Beard Grass   181
Golden Bristle Grass   94
Golden Velvet Grass   151
Gonya Grass   160
Goose Grass   157
Guinea Grass   239
Gum Grass   189

H

Hairy Blue Grass   124
Hairy Love Grass   232
Hairy Trident Grass   208
Hare's Tail   111
Harpoon Grass   195
Heart-seed Love Grass   188
Herringbone Grass   199
Hippo Grass   159
Horse Grass   93

I

Indian Goose Grass   157
Iron Grass   214

J

Jungle Rice   170

K

Kalahari Sand Quick   201
Kalahari Sour Grass   200
Kikuyu   280
Knee Grass   226

L

L.M. Grass   126
Land Grass   241
Large Carrot-seed Grass   100
Large Panicle Agrostis   210
Large Scale Grass   169
Large-seeded Three-awn   180
Large Silver Andropogon   266
Lehmann's Love Grass   226
Little Quaking Grass   219
Long-awned Three-awn   105, 106
Love Grass
    Bronze Love Grass   225
    Broom Love Grass   191
    Cape Love Grass   188
    Elastic Love Grass   231
    False Love Grass   162
    Flat-seed Love Grass   194
    Footpath Love Grass   229
    Hairy Love Grass   232
    Heart-seed Love Grass   188
    Lehmann's Love Grass   226
    Narrow Heart Love Grass   193
    Rough Love Grass   221
    Sawtooth Love Grass   194
    Shade Love Grass   222
    Sticky Love Grass   233
    Sticky-stem Love Grass   189
    Tough Love Grass   192
    Viscid Love Grass   233
    Weeping Love Grass   224
    Wether Love Grass   190
    Woolly Love Grass   109

M

Marsh Grass   170
Mountain Bristle Grass   114
Mountain Wire Grass   83
Mouse Barley   80
Munnik Fescue   234

N

Narrow Heart Love Grass   193
Narrow-leaved Turpentine Grass   268
Natal Crowfoot   126
Natal Panicum   240
Natal Red Top   235
Needle Grass   119
Ngongoni Three-awn   104
Nile Grass   161
Nine-awned Grass   186

O

Oat Grass
    Small Oat Grass   110
One-finger Grass   118

P

Pale Three-awn   179
Pan Dropseed   247
Panicum
    Broad-leaved Panicum   237
    Natal Panicum   240
    Plum Panicum   238
    Reed Panicum   237

Small Panicum   236
Paspalum
   Buffalo Quick Paspalum   131
   Common Paspalum   173
   Couch Paspalum   131
   Giant Paspalum   260
   Tall Paspalum   260
   Veld Paspalum   174
Perennial Rye Grass   82
Perennial Signal Grass   160
Perennial Spiderweb Grass   136
Pinchushion Grass   121
Pinhole Grass   153
Plum Panicum   238
Purple Finger Grass   149
Purple Plume Grass   217
Purple Spike Grass   87
Purple Three-awn   215

Q

Quaking Grass
   Big Quaking Grass   256
   Little Quaking Grass   219
Quick Grass
   Common Quick Grass   139
   Sand Quick   201
Quinine Grass   101

R

Rainbow Vlei Grass   156
Ratstail Dropseed   116
Red Autumn Grass   89
Red Thatching Grass   272
Red Top
   Bristle-leaved Red Top   197
   Mountain Red Top   197
   Natal Red Top   235
Red Vlei Grass   159
Rescue Grass   257
Rhodes Grass   135
   Brown Rhodes Grass   152
Ripgut Brome   258
River Grass   107
Robies Cocksfoot   175
Rolling Grass   212, *242*
   Small Rolling Grass   253
Rooigras   278
Rough Love Grass   221
Russet Grass
   Common Russet Grass   196

S

Sand Quick   201
Sawtooth Love Grass   194
Setaria
   Black-seed Setaria   92
   Broad-leaved Setaria   261
   Common Setaria   95
   Creeping Setaria   96
   Fan-leaved Setaria   202
   Garden Setaria   93
   Golden Setaria   94
   Mountain Setaria   114
   Shade Setaria   97
Shade Bristle Grass   97
Shade Love Grass   222
Sickle Grass   117
Sign Grass   127
Signal Grass
   Black-footed Signal Grass   166
   Bushveld Signal Grass   176
   Common Signal Grass   176
   False Signal Grass   218
   Garden Signal Grass   177
   Perennial Signal Grass   160
   Sweet Signal Grass   164
   Velvet Signal Grass   167
Silky Bushman Grass   251
Silky Autumn Grass   88
Silver Finger Grass   143
Silver Thread Grass   264
Silver Wool Grass   74
Silverspike   81
Small Bushman Grass   205
Small Creeping Foxtail   96
Small Oat Grass   110
Small Rolling Grass   253
Small Scale Grass   168
Smut's Finger Grass   145
Snowflake Grass   264
Sour Grass   77, 200
   Grey Sour Grass   186
Spear Grass   79
   Giant Spear Grass   98
Spiderweb Grass   137
   Perennial Spiderweb Grass   136
Spreading Three-awn   213
St. Augustine Grass   122
Stab Grass   125
Staggers Grass   112
Star Grass   140
Sticky Love Grass   233
Stinking Grass   154
Swamp Grass   184
Sweet Buffalo Grass   241
Sweet Signal Grass   164

T

Tall Bushman Grass   204
Tall Three-awn   255
Tanglehead   79
Tassel Grass   198
    Purple Tassel Grass   217
Tassel Three-awn   103
Teff   259
Thatching Grass
    Blue Thatching Grass   275
    Boat Thatching Grass   271
    Bundle Thatching Grass   270
    Common Thatching Grass   274
    Fine Thatching Grass   273
    Red Thatching Grass   272
    Yellow Thatching Grass   276
Thimble Grass   78
Thread-leaved Bluestem   130
Three-awn
    Annual Three-awn   102
    Coppery Three-awn   254
    Giant Three-awn   254
    Large-seeded Three-awn   180
    Long-awned Three-awn   105, 106
    Ngongoni Three-awn   104
    Pale Three-awn   179
    Purple Three-awn   215
    Rough Three-awn   180
    Spreading Three-awn   213
    Tall Three-awn   255
    Tassel Three-awn   103
Three-awned Rolling Grass   212
Tick Grass   171
Tough Love Grass   192
Trident Grass   208
Triraphis   207
Tumble Weed   242
Turf Grass   158
    Yellow Turf Grass   90
Turpentine Grass
    Bitter Turpentine Grass   268
    Broad-leaved Turpentine Grass   267
    Common Turpentine Grass   267
    Giant Turpentine Grass   269
    Narrow-leaved Turpentine Grass   268

V

Vasey Grass   260
Veld Paspalum   174
Velvet Grass
    Golden Velvet Grass   151
Velvet Signal Grass   167
Vlei Bluestem   134
Vlei Bristle Grass   91
Vlei Finger Grass   155
Vlei Grass   263
    Catstail Vlei Grass   183
    Rainbow Vlei Grass   156
    Red Vlei Grass   159

W

Weeping Love Grass   224
Wether Love Grass   190
White Buffalo Grass   236
Wild Oat
    Common Wild Oat   216
    Wild Oat Grass   277
Wild Sorghum
    Black-seed Wild Sorghum   262
    Common Wild Sorghum   245
Winter Grass   243
Wire Grass   77, *104*
    Mountain Wire Grass   83
Wool Grass   75
    Silver Wool Grass   74
Woolly Love Grass   109

Y

Yellow Thatching Grass   276
Yellow Turf Grass   90

ACKNOWLEDGEMENTS

We thank the following companies for their sponsorship and donations.

The Anglo American and De Beers Chairman's Fund

KYNOCH KUNSMIS   **SENSAKO**